PRACTICAL
PLANT BIOCHEMISTRY

PRACTICAL
PLANT BIOCHEMISTRY

BY

MURIEL WHELDALE ONSLOW, M.A.

LECTURER IN PLANT BIOCHEMISTRY, UNIVERSITY OF CAMBRIDGE
AUTHOR OF *THE ANTHOCYANIN PIGMENTS OF PLANTS*

CAMBRIDGE
AT THE UNIVERSITY PRESS
1929

CAMBRIDGE
UNIVERSITY PRESS

University Printing House, Cambridge CB2 8BS, United Kingdom

Published in the United States of America by Cambridge University Press, New York

Cambridge University Press is part of the University of Cambridge.

It furthers the University's mission by disseminating knowledge in the pursuit of education, learning and research at the highest international levels of excellence.

www.cambridge.org
Information on this title: www.cambridge.org/9781107634312

© Cambridge University Press 1929

First edition 1920
Second edition 1923
Third edition 1929
First published 1929
First paperback edition 2014

A catalogue record for this publication is available from the British Library

ISBN 978-1-107-63431-2 Paperback

Cambridge University Press has no responsibility for the persistence or accuracy of URLs for external or third-party internet websites referred to in this publication, and does not guarantee that any content on such websites is, or will remain, accurate or appropriate.

PREFACE

THIS book is intended primarily for students of Botany. Such a student's knowledge of plant products is usually obtained, on the one hand, from Organic Chemistry, on the other hand, from Plant Physiology; between these two standpoints there is a gap, which, it is hoped, the following pages may help to fill. It is essentially a text-book for practical work, an aspect of Plant Biochemistry which has received up to the present time very little consideration in teaching. A number of experiments have been devised and have been actually tested in practical classes. These experiments should enable a student to extract from the plant itself the chemical compounds of which it is constituted, and to learn something of their properties. An elementary knowledge of Organic Chemistry on the part of the student has been assumed, as it appeared superfluous to incorporate the material which has already been so amply presented in innumerable text-books.

My sincerest thanks are due to Dr F. F Blackman, F.R.S., for criticism and many suggestions throughout the writing of the book. I am further indebted to Mr H. Raistrick, M.A., for help in various ways, especially in reading the proof-sheets. I wish, in addition, to express my gratitude to Professor F. G. Hopkins, F.R.S., for the great interest he has always shown in the subject and for his kind and stimulating advice in connexion with the scheme of teaching presented in the following pages.

M. W. O.

CAMBRIDGE,
February, 1920.

PREFACE TO THE SECOND EDITION

IN the present edition, some account, accompanied in most cases by illustrative experiments, has been given of a number of substances, or groups of substances, involved in plant metabolism, which were not included in the first edition. These are notably the "vegetable acids," waxes, sterols, lecithins, inositol, phytin, the "essential oils" and nucleic acid. Corrections have also been made in order to include more recent additions to our knowledge on certain problems, as, for instance, those connected with oxidizing enzymes.

Since it is advisable to keep the book as short as possible, a few of the original experiments have been omitted to make space for others considered to be of greater value to the student.

The chapter on the colloidal state is intended to give the student a preliminary conception, only, of the importance of such phenomena. Additional information, both as to theory and experiment, is to be found in text-books which deal more exclusively with this subject.

Sufficient experience has not yet been gained to admit of the inclusion, in the present edition, of quantitative class-work in Plant Biochemistry.

I am much indebted to Dr F. F. Blackman, F.R.S., for kindly assisting with the proofs.

M. W. O.

CAMBRIDGE,
December, 1922.

CONTENTS

CHAPTER I

INTRODUCTION

This chapter should be re-read after the remaining chapters have been studied.

ALL plants are made up of a complex organized mixture of chemical substances, both organic and inorganic. As a preliminary to the study of plant chemistry, the student should realize that the chemical compounds which make up the living plant may be approximately grouped into the six following classes. Thus, in later chapters, when reference is made to any plant product, it will be understood, broadly speaking, to which class it belongs, and what relationship it bears to other chemical compounds.

The main classes may be enumerated as follows:

(1) *Carbohydrates.* The simplest members of this class are the sugars, which are aldehydes and ketones of polyhydric alcohols of the methane series of hydrocarbons. The more complex carbohydrates, such as starch, cellulose, dextrins, gums and mucilages, are condensation products of the simpler sugars. The sugars are found in solution in the cell-sap of living cells throughout the plant. Cellulose, in the form of cell-walls, constitutes an important part of the structure of the plant, and starch is one of the most widely distributed solid "reserve materials."

(2) *Vegetable acids.* This term is usually applied to acids and hydroxy-acids derived from the *lower* members of the methane, olefine and acetylene series of hydrocarbons. Such acids as formic, acetic, valeric and caproic are not readily detected in the plant. Nevertheless, it is more than likely that they play an important part in metabolism, for their amino derivatives, glycine, valine, etc. (see section 5) form constituents of practically all proteins. The dibasic and hydroxy-acids, e.g. oxalic, succinic, glutaric, malic, etc., are probably products of oxidation of the sugars in respiration. Aspartic (amino-succinic) and glutaminic (amino-glutaric) acids are also constituents of proteins.

(3) *Fats.* Chemically these are glycerides, that is glycerol esters, of acids derived from the *higher* members of the methane and olefine series of hydrocarbons, and they usually contain a large number of carbon atoms. The fats occur as very fine globules deposited in the cells, especially in the tissues of seeds where they form reserve materials, though they also occur in other parts of plants.

o.

The lecithins, which are compounds of fats with phosphoric acid, are probably present in all living cells and have an important metabolic significance.

The above substances belong to the aliphatic series of organic compounds, that is to the series in which the carbon atoms are united in chains.

(4) *Aromatic compounds.* These are characterized by having the carbon atoms united in a ring as in benzene. They may contain more than one carbon ring, and, moreover, aliphatic groupings may be attached to the carbon ring as side-chains. The number of aromatic substances is very great, and every plant contains representatives of the class. Some members are widely distributed; others, as far as we know, are restricted in their distribution, and may be peculiar to an order, a genus or even a species. This class contains: (*a*) Phenols, i.e. hydroxy-derivatives of benzene, such as phloroglucinol. (*b*) Aromatic alcohols, aldehydes and acids derived from benzene; various hydroxy-benzoic acids, such as gallic and protocatechuic acids, are important, since, by condensation, they give rise to tannins. Just as in the case of the carbohydrates, where simpler compounds may become more complex by condensation, the soluble crystalline acids condense to form the complex colloidal tannins. Of other aromatic acids, the amino derivatives, such as phenylalanine and tyrosine, form constituents of proteins. (*c*) Complex hydrocarbons, the terpenes, accompanied by derivative alcohols, aldehydes, ketones and esters. These form constituents of the "essential oils" obtained from plants by steam distillation, and are responsible for most of the plant scents. (*d*) Other members which contain more than one ring are the water-soluble yellow, red, purple and blue pigments of plants, the yellow being hydroxy-flavones and flavonols, the remainder, anthocyan pigments.

(5) *Proteins.* This large class contains substances which are in many cases built up of groupings from both the aliphatic and aromatic series. It includes not only the proteins but also their simpler derivatives, the albumoses, peptones and polypeptides. In this case, as before, the simplest derivatives, known as the amino-acids, are synthesized by condensation to form the polypeptides, peptones, albumoses and proteins, in a series of increasing complexity. The amino-acids are compounds, either of the aliphatic, aromatic or heterocyclic (see 6) series, in which one or more hydrogen atoms are replaced by the radicle NH_2. They are soluble and crystalline, but after condensing together, the final product, the protein, only exists in either the solid or the colloidal state. Proteins, in the latter condition, constitute the bulk of the complex material, protoplasm;

in the solid state, in the form of grains and granules, they occur as reserve material in the cell.

(6) *Plant bases.* This class contains (*a*) the amines or substitution products of ammonia. Sometimes the hydrogen of ammonia is substituted by a group of some complexity which leads to the production of a compound of the heterocyclic type, i.e. with a ring containing both carbon and nitrogen atoms. The pyrrole ring is an example which occurs in the amino-acid, proline, in certain alkaloids (see below), and in the pigment chlorophyll. (*b*) Purines. In connection with these substances we need to consider two more heterocyclic rings, i.e. the pyrimidine and the iminazole. The former may be regarded as the condensation product of urea, which is possibly present in small quantities in plants, and an unsaturated acid, e.g. acrylic acid. The pyrimidine ring is present in some purines, the iminazole in the amino-acid, histidine. The remaining purines contain a condensed pyrimidine and iminazole ring. Certain of the purines become condensed together, in combination with phosphoric acid and a pentose sugar, to form the nucleic acids. The latter, in combination with proteins, as nucleoproteins, form a constituent, as their name implies, of the nucleus. (*c*) The alkaloids are substances of considerable complexity, containing various heterocyclic rings. Unlike the simpler bases, they are restricted to a certain extent in their distribution.

It is not possible to include all classes of plant substances in the above list and many others, such as the sulphur compounds, sterols, phytin, etc., are referred to in the later chapters. It should be borne in mind that the importance of a compound in plant metabolism is not estimated by the amount of it occuring in the plant. Frequently, most important substances occur in such small quantities that they are difficult to detect.

In order to appreciate the subject of plant chemistry, the plant, which is familiar as a botanical entity, must be interpreted in chemical terms. The principal classes of the more essential and widely distributed compounds found in plants have already been indicated on the broadest basis, so that they may now be referred to without additional comment.

From the botanical point of view, the plant may be regarded as a structure composed of many living protoplasmic units enclosed in cell-walls and combined together to form tissues. There are also certain tissues, known as dead tissues, which assist in giving rigidity to the plant. All these structural elements may, in time, be translated into terms of chemical compounds.

One of the chemical processes most frequently met with in the plant is that of synthesis by condensation, with elimination of water, of large complex molecules from smaller and simpler molecules. The formation of cellulose, for instance, is a case in point. Cellulose has the composition $(C_6H_{10}O_5)_n$ and, on hydrolysis with dilute acids, it yields glucose as a final product. Hence it is concluded that the complex molecule of cellulose is built up from the simpler carbohydrate by condensation. The synthesis of proteins from amino-acids affords another example. These acids contain either an aliphatic or aromatic nucleus (let it be R), and one or more carboxyl and amino groups. Condensation takes place in the plant, with elimination of water, according to the following scheme:

$$\underset{\text{NH}_2\cdot\text{CH}-\text{CO}}{\overset{R^i}{|}} \text{OH} \quad H \underset{\text{NH}\cdot\text{CH}-\text{CO}}{\overset{R^{ii}}{|}} \text{OH} \quad H \underset{\text{NH}\cdot\text{CH}-\text{CO}}{\overset{R^{iii}}{|}} \text{OH} \ldots\ldots H \underset{\text{NH}\cdot\text{CH}-\text{COOH}}{\overset{R^z}{|}}$$

The products of such condensation, the proteins, vary among themselves according to the number and kind of amino-acids which take part in the synthesis.

Two important results arise from this process. First, the substances formed by condensation have molecules of a very large size ; secondly, whereas the simple compounds, sugars and amino-acids, are soluble, crystalline and diffusible, the condensation products are either insoluble, e.g. cellulose, or exist in the colloidal state, as is the case of many proteins and other plant constituents. As these very large molecules do not dialyze, they remain where they are synthesized, and build up the solid structure of the plant, as for instance, the cell-walls.

Matter in the colloidal state is of very great importance in the plant and is probably responsible for many of the properties of "living" material. Thus it will not be out of place, though it will be referred to again in a later chapter, to make at this point a few remarks on the colloidal state. It has been known for some time that certain metals, e.g. gold and silver, and also certain metallic hydroxides and sulphides, e.g. ferric hydroxide and arsenious sulphide, though insoluble in water under ordinary conditions, can, by special methods, be obtained as solutions which are clear to the unaided vision. Such solutions are termed colloidal. Investigation has shown that the matter is not present in true solution, but in a very finely divided state, i.e. as particles many times larger than simple molecules, but smaller than the particles obtainable by mechanical means of division. Such solutions are known as artificial colloidal solutions, but there are a number of organic substances, with *very* large molecules, such as proteins, starch, gums, agar, etc., which at once dissolve in water giving colloidal solutions. The main feature of the colloidal state is that the system consists of two phases, or conditions of matter. In the case of the artificial colloidal solutions first mentioned, one state is solid, the gold particles; the other state is liquid, the water. The solid is known as the dispersed phase, and the water as the continuous phase, and such colloidal solutions are termed suspensoids. In the case of proteins, starch, etc., both phases are liquid: the dispersed phase, a concentrated solution of protein, etc.; the continuous phase, a dilute solution of protein, etc. Such colloidal solutions are known as emulsoids.

An important point in connexion with the colloidal state is that the molecules, or aggregates of molecules, forming the dispersed phase are so large that they exhibit some of the phenomena of surface energy, electrical charge, etc., associated with matter in mass. These properties come to be of considerable importance, when we consider how large a surface is presented by matter in this state in comparison with its mass.

A material in the plant upon which much interest naturally centres is the protoplasm and the nucleus. It has been shown that the protoplasm consists, chemically, largely of proteins in the colloidal state. It is itself a liquid, and embedded in it are substances of various chemical constitution, in the form of granules of solid matter and also liquid globules. Numerous chemical reactions are continually taking place in the protoplasm throughout the cell, and since many of these reactions can take place both simultaneously and independently, the protoplasm must have some form of organized structure. Though many phenomena of "life" may be accounted for by the physical and chemical properties of such substances as proteins, it is impossible to say, with our present knowledge, how far all "living" phenomena may yet be explained in this way.

Some of the main lines of metabolic syntheses which take place in the plant will next be considered. A fundamental fact which should be borne in mind is that the green plant synthesizes all the complex materials of which it is composed from the simple compounds, carbon dioxide, water and certain inorganic salts. The most important factor, perhaps, which figures in plant metabolism, is chlorophyll. The green pigments of chlorophyll are esters of complex organic acids containing the elements carbon, hydrogen, oxygen, nitrogen and magnesium. They have the remarkable power of absorbing the radiant energy of the sun's rays and of transforming it into chemical energy, by means of which carbon dioxide and water are combined to form some organic compound, possibly formaldehyde, from which a simple carbohydrate is readily synthesized.

If now the initial and final products of carbon assimilation be considered in detail, it will be seen that the process is one of reduction :

$$6CO_2 + 6H_2O = C_6H_{12}O_6 + 6O_2.$$

This is confirmed by the fact that oxygen is evolved in the process. Moreover, the plant accumulates a store of energy, since the final product, the carbohydrate, has a higher potential energy than the system, water and carbon dioxide. Hence carbon assimilation, in addition to providing a basis of organic material as a starting-point for all the main metabolic functions, also provides a source of chemical energy by means of which reactions in other directions are brought about.

The setting free of this accumulated energy constitutes the process of respiration, which is, in reality, an oxidation of carbohydrate taking place in tissues throughout the plant. It is the converse of carbon assimilation, in that oxygen is absorbed and carbon dioxide and water are formed. Thus these two processes, both so fundamental and essential to the metabolism of the green plant, are constantly taking place side by side in the same cell.

The first-formed carbohydrate, which is probably a hexose, is condensed in the plant, on the general lines we have previously indicated, to form more complex disaccharides and polysaccharides, such as maltose, cane-sugar, starch, cellulose, etc. Some of these products, such as the disaccharides, form true solutions and may be present in the cell-sap; others, such as cellulose and starch, are present in the solid state, though they contain considerable quantities of water. Others, again, such as dextrin and gum, are present in the colloidal state. Thus, given an initial carbohydrate and a source of energy, we may proceed to indicate the other main lines of syntheses in the plant.

The next most important line of syntheses is probably that which gives rise to the nitrogen-containing constituents of the plant. Nitrogen is absorbed by the green plant in the form of nitrates and ammonium salts, but the processes which lead to the synthesis of some of the simplest nitrogen-containing compounds, such as the amino-acids, are still very obscure. Aliphatic and aromatic acids of various kinds are abundantly present in the tissues, but the reactions by which the NH_2 groups are introduced are by no means clear. There is little doubt, however, that once the amino-acids are formed, condensation takes place as already indicated, and more complex molecules, termed polypeptides, arise. Such polypeptides have now been synthesized artificially by the condensation of amino-acids. From the polypeptides, by further stages of condensation, the albumoses, peptones, and finally proteins are produced.

Another line of syntheses is that which leads to the production of the fats and allied substances. The fats are mainly glycerides of acids of the methane and olefine series, such as butyric, palmitic and oleic acids. Like all other plant products the fats must either directly or indirectly arise from the carbohydrates. There is evidence that the origin is fairly direct, as, for instance, in fatty seeds when the fats take the place of sugars in ripening. The sugars, as we know, are aldehydes of the polyhydric alcohols of the methane series. It has been suggested, though the actual stages have not been ascertained, that by various oxidation and reduction processes, the sugars yield fatty acid residues which then condense to form the fatty acids of high molecular weights present in fats. By a converse process, the fats, especially when they are stored as reserve materials in seeds, are broken up, and sugars are again formed which pass to other parts of the germinating seedling, and are there used in other synthetic processes.

A third main line of syntheses is that which gives rise to the aromatics of the plant. Since no ring compound is absorbed by the green plant, it follows that by some process the aliphatic structure must be transformed into the aromatic. Thus, for instance, the trihydric phenol, phloroglucinol, might at some stage be formed from a hexose by conversion of the aliphatic chain into a closed ring:

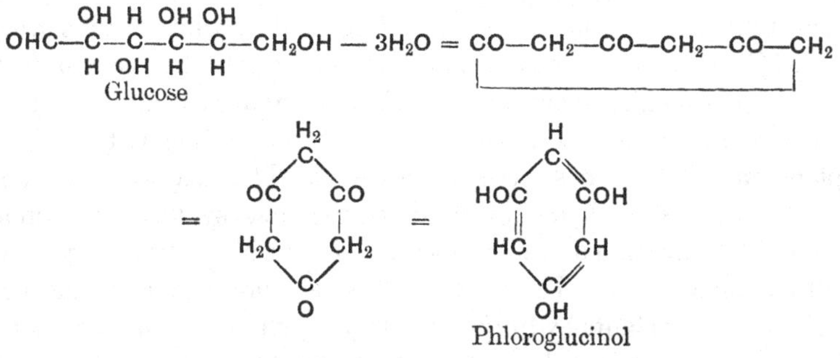

$$OHC-\overset{\overset{OH}{|}}{\underset{\underset{H}{|}}{C}}-\overset{\overset{H}{|}}{\underset{\underset{OH}{|}}{C}}-\overset{\overset{OH}{|}}{\underset{\underset{H}{|}}{C}}-\overset{\overset{OH}{|}}{\underset{\underset{H}{|}}{C}}-CH_2OH \ - \ 3H_2O \ = \ CO-CH_2-CO-CH_2-CO-CH_2$$

Glucose

$$= \quad \begin{matrix} & H_2 \\ & C \\ OC & & CO \\ | & & | \\ H_2C & & CH_2 \\ & C \\ & O \end{matrix} \quad = \quad \begin{matrix} & H \\ & C \\ HOC & & COH \\ \| & & | \\ HC & & CH \\ & C \\ & OH \end{matrix}$$

Phloroglucinol

There is evidence that aromatic compounds, such as phloroglucinol, tannins, flavones and anthocyanins are synthesized in the leaves, and that sugar-feeding, by floating leaves in sugar solutions, leads to the increase of aromatics in the tissues. When the ring structure has been once synthesized, further changes can take place either by the addition of side-chains to the ring or by the condensation of two or more rings. In this way the great multitude of aromatic products present in the higher plants may arise.

Thus the cell can be pictured as a colloidal solution of proteins endowed with the properties of matter in mass and surrounded by a permeable cell-wall of cellulose. The colloidal solution contains liquid and solid particles of very varied chemical composition. In the protoplasm are spaces, vacuoles, filled with cell-sap also containing many and various substances in solution. Throughout the protoplasm, which probably has an organized structure, many kinds of chemical reactions are continually in progress, some being the converse of others, as for instance those of oxidation and reduction which can take place side by side in the same cell.

Next will be considered the chemical reactions by which the various metabolic changes in the plant are brought about. How are these processes controlled and how do they take place?

There is a large group of organic substances, termed enzymes, many of which are present in every plant. They have a certain characteristic in common, i.e. they bring about chemical reactions in the plant without undergoing any permanent change: in other words they are organic catalysts. Many of these reactions, which take place in the cell at ordinary temperatures with considerable rapidity, need prolonged heating at high temperatures when brought about by artificial means. Enzymes can generally be extracted from the plant by water, especially if the tissues are thoroughly disintegrated. Their chemical constitution is at present unknown, and they are usually destroyed by temperatures greater than 60° C. Moreover, many of the processes which they control in the plant can be brought about by them *in vitro* under suitable conditions, and it is by means of such experiments that information as to their rôle in plant metabolism has been ascertained. The majority of known enzymes control both hydrolysis and its converse, synthesis by condensation with elimination of water, but under artificial conditions hydrolysis most frequently occurs. The enzyme, diastase, for instance, found in all starch-containing plants hydrolyzes *in vitro* starch to dextrin and maltose. Similarly the enzyme, maltase, hydrolyzes maltose into glucose. Other enzymes hydrolyze proteins into amino-acids, and others, again, hydrolyze fats into fatty acids and glycerol.

Until fairly recently the fact escaped notice that such reactions are reversible, and that these enzymes *in situ* in the plant may, according to the conditions, control not only the hydrolytic but also the corresponding synthetic process. The latter may also be brought about, though not readily, *in vitro*. This, and other evidence, leads us to believe that enzymes in the plant control the reactions in both directions.

Hydrolysis, and synthesis with elimination of water are not however the only processes catalyzed by enzymes. There is another type of these catalysts, the oxidizing enzymes, which bring about oxidation of substances in the plant, notably of aromatics. In addition, there is the enzyme, zymase, which decomposes sugar with the production of alcohol and carbon dioxide.

The question which now arises is—How many reactions in the plant are catalyzed by enzymes? It is conceivable that a greater number of enzymes may exist than are at present known, but that they are unable to be extracted by our present methods of isolation. A certain number of reactions probably take place in the cell-sap between the substances in solution; others are catalyzed by enzymes which are supposed to be intimately connected with the protoplasm, but there are an enormous number to which there is at present no clue as to how they are brought about, such, for instance, as the synthesis of carbohydrates from carbon dioxide and water, and the formation of the benzene ring from the open carbon chain. Such processes are usually said to be controlled by the "living protoplasm," but what exactly is the significance of this expression is at present beyond our knowledge.

Finally, also, little is known of the question as to how the various lines of metabolic syntheses in different parts of plants are regulated and correlated with each other. Some of the phenomena involved are shortly outlined as follows. There is undoubtedly, under suitable conditions, a constant synthesis of sugars in the leaves. In all probability aromatic substances are also synthesized in the same organs, for there is evidence that there is an increase of these compounds in the leaf if translocation through the petiole is prevented. It is possible that amino-acids also are formed in the leaf. The above products are constantly translocated to the growing organs as material for growth. They may, nevertheless, be temporarily stored in the tissues where they have been synthesized, and of this there is evidence in at least one case, e.g. starch in the leaf. But, apart from the immediate use for growth, there is in practically every plant, some tissue where, owing to some unknown stimulus (causing probably changes in permeability of the cell-membranes), accumulation of compounds occurs. This accumulation is characteristic of organs from which growth will take place when it is impossible for the plant to obtain fresh supplies by carbon assimilation, as, for example, of bulbs, rhizomes, tubers, buds, seeds, fruits and woody tissues. In these cases, in due time, the products stored supply the growing shoots.

During storage, simple sugars, amino-acids, etc. have been condensed to form insoluble, colloidal, or large molecules of starch, fats, aleurone, cane-sugar, etc. These will remain until they are hydrolyzed by enzymes when they can supply the growing shoots. Such stores are termed "reserve materials." The actual stimuli involved in bringing about and regulating this storage are unknown, but they are probably connected with the life cycle of the particular plant under consideration and its adaptation to external conditions.

REFERENCES

1. **Abderhalden, E.** Handbuch der biochemischen Arbeitsmethoden. Berlin, 1920- .

2. **Abderhalden, E.** Biochemisches Handlexikon. Berlin, 1911.

3. **Allen's** Commercial Organic Analysis. London, 1924–28. 5th ed.

4. **Bertrand, G.,** and **Thomas, P.** Practical Biological Chemistry. Translated by H. A. Colwell. London, 1920.

5. **Cole, S. W.** Practical Physiological Chemistry. Cambridge, 1928. 8th ed.

6. **Czapek, F.** Biochemie der Pflanzen. Jena, Bd. 1, 1913, Bd. 2, 1920, Bd. 3, 1921.

7. **Haas, P.,** and **Hill, T. G.** The Chemistry of Plant Products. London, 1928. 4th ed.

8. **Kostytschew, S.** Lehrbuch der Pflanzenphysiologie. Bd. 1. Chemische Physiologie. Berlin, 1926.

9. **Kostytschew, S.** Pflanzenatmung. Berlin, 1924.

10. **Palladin, V. I.** Plant Physiology. Edited by B. E. Livingston. Philadelphia, 1918.

11. **Plimmer, R. H. A.** Practical Organic and Biochemistry. London, 1926. 4th ed.

12. **Rosenthaler, L.** Grundzüge der chemischen Pflanzenuntersuchung. Berlin, 1923.

13. **Wehmer, C.** Die Pflanzenstoffe. Jena, 1911.

14. **Wester, D. H.** Anleitung zur Darstellung phytochemischer Uebungspräparate. Berlin, 1913.

CHAPTER II

THE COLLOIDAL STATE

MANY of the substances of which the plant is built up exist in the living cell in the colloidal state, and it is therefore important that some account should be given of this condition of matter.

There are many organic products found in the plant (and also in the animal), such as starch, various proteins, gums, etc., that apparently dissolve in water, giving a solution which, as a rule, only differs from an ordinary solution by being opalescent. In addition, it has been known for a long time that various inorganic substances, such as sulphides of arsenic and antimony, hydroxide of iron, and also certain metals (gold, silver), can, by special methods, be obtained in "solution," though in ordinary circumstances they are quite insoluble. The above examples are representative of colloidal solutions.

A property which all the above solutions possess is that the substance dissolved will not pass through a parchment membrane, i.e. will not dialyze, whereas if a solution of sodium chloride in water is separated from pure water by a parchment membrane, the salt will pass through the membrane until the concentration of the sodium chloride is equal on either side of it.

The conclusion drawn from investigations of various kinds is that in the colloidal solutions the substances dissolved exist in the state, either of aggregates of molecules, or of very large molecules, and hence are unable to pass through the pores of the parchment.

Moreover, certain distinctions can be drawn between colloidal solutions: some, like those of gold, silver, metallic sulphides, hydroxides and in fact most inorganic substances, are very sensitive to the presence of small amounts of inorganic salts, i.e. electrolytes, and are precipitated by them, but will not as a rule go into solution again. Also such colloidal solutions are very little more viscous than pure water. The organic substances, on the other hand, are only precipitated from colloidal solutions by comparatively large quantities of electrolytes. The viscosity, moreover, of these solutions is greater than that of water, and is, in fact, considerable, even if the percentage of dissolved matter is small.

Hence two terms have been employed for the above-mentioned types of colloidal solutions: those of gold, silver, etc., are termed *suspensoids* (suspensoid sols): those of starch, proteins, etc., *emulsoids* (emulsoid sols).

The essential feature of both forms is that they are systems consisting of two phases, or conditions of matter, known respectively as the "dispersed" phase and the "continuous" phase.

A *suspensoid* may be defined as having a *dispersed phase* composed of ultramicroscopic particles or aggregates of molecules suspended in a *continuous phase* composed of a liquid.

An *emulsoid* may be defined as having a *dispersed phase* composed of ultramicroscopic drops of a highly concentrated solution of the substance suspended in a *continuous phase* composed of a dilute solution of the same substance.

As a rule, therefore, the difference between a suspensoid and an emulsoid is that, whereas in the former the liquid is restricted to the continuous phase, and the solid to the dispersed phase, in an emulsoid both phases are liquid, though containing different proportions of the dissolved substance.

The terms suspensoid and emulsoid are used on account of the resemblance of these states of matter respectively to suspensions and emulsions. If microscopic particles of a solid are shaken up in water, what is known as a suspension is obtained; in time, however, the solid particles, if heavy enough, will settle and separate from the water, and the whole process can be repeated. Thus a suspension differs from a suspensoid solution in that the latter is stable, though, if precipitated, the reaction is usually not reversible.

If two liquids which are insoluble in each other, such as oil and water, are shaken up together, finely divided drops of oil in water are obtained. This is known as an emulsion. In time, however, the oil separates from the water, because the tension on the films of water separating the oil drops, when in contact, is too great, and they break, with the result that the oil drops coalesce. But if, instead of water, a solution of soap, saponins, or certain other substances is used, the surface tension of the water is so lowered that the films of soap solution separating the oil drops are permanent, and a system is obtained consisting of minute drops of oil separated by soap solution. This system resembles an organic colloidal solution, as, for instance, that of protein in which we suppose a concentrated solution of protein exists in drops separated by a dilute solution of protein. Milk and latex constitute natural emulsions.

Expt. 1. *Formation of a suspension.* Precipitate a solution of barium chloride with some sulphuric acid and shake up well the fine precipitate of barium sulphate. Note the gradual settling of the precipitate.

Expt. 2. *Formation of an emulsion.* Take a drop of olive oil in a test-tube and half fill the tube with alcohol. Shake well and pour into a beaker of water. A fine white emulsion of oil in water will be formed from which the oil will not separate. By this method the oil is obtained in such small drops that stability is ensured.

Take about equal quantities of olive oil in two test-tubes and add an equal quantity of water to each. To one tube add a drop or two of 10 % caustic alkali solution. Shake both test-tubes well. An emulsion is formed in both, but in the tube without alkali the oil will separate out on standing. In the other tube the emulsion is permanent. This is due to the fact that the olive oil (unless specially purified) contains some free fatty acid. The latter forms soap with the alkali (see p. 93) and renders the emulsion permanent.

Expt. 3. *Preparation of suspensoid sols.* (a) *Gold.* Take two 100 c.c. measuring cylinders and thoroughly clean them with nitric acid, and afterwards wash well with freshly distilled water. In one make a 0·5 % solution of tannic acid (using the purest sample obtainable) in water. In the other, 2 c.c. of commercial 1 % gold chloride are made up to 100 c.c. with water. Use freshly distilled water in both cases. Mix equal portions of the two solutions in a clean beaker. A purple colloidal solution of gold will be formed. If three parts of the chloride solution are mixed with one part of the tannin solution, and both solutions heated before mixing, a red colloidal solution is obtained. (b) *Silver.* Take 5 c.c. of a 1 % solution of silver nitrate and add dilute ammonia solution until the precipitate first formed just disappears, and then dilute with 100 c.c. of water. Mix equal volumes of this solution and the tannic acid prepared for (a). A colloidal solution of silver will be formed which is clear brown by transmitted light, but has a green fluorescence by reflected light. (c) *Ferric hydroxide.* Take 5 c.c. of a filtered 33 % solution of ferric chloride and pour into 500 c.c. of boiling distilled water in a beaker. A colloidal ferric hydroxide sol is formed and the colour changes to a deep brown-red. The yellow solution of ferric chloride is decomposed by excess of water with the production of a soluble colloidal form of ferric hydroxide, and hydrochloric acid is set free. (d) *Arsenic trisulphide.* Take 2 gms. of arsenious acid and boil with 150 c.c. of distilled water, filter and cool. Then pass sulphuretted hydrogen through the solution. A colloidal solution of the sulphide is formed which is orange, with a greenish surface.

The above sols should be kept for further experiment [see Expt. 8].

Expt. 4. *Preparation of emulsoid sols.* (a) *Starch.* Weigh out 2 gms. of dry starch, and mix well with a little cold distilled water. Boil 100 c.c. of distilled water in a flask, and, when boiling, pour in the starch paste and boil for a few minutes longer, stirring well all the time. A colloidal solution of starch is obtained which is faintly opalescent. It is not affected by heating and does not change its state on cooling. (b) *Gum arabic.* Make a 5 % solution of gum arabic by boiling 5 gms. with 100 c.c. of distilled water. Note that a sticky or viscous solution is formed which froths on shaking. (c) *Protein.* Weigh out 10 gms. of white flour and add 100 c.c. of distilled water. Let the mixture stand for 2 or 3 hours and then filter. The extract contains protein. Note that the solution froths on shaking. (*d*) *Soap*. Make a 5-10 % solution of soap in distilled water. It is opalescent and froths strongly.

The above sols should be kept for further experiment [see Expt. 9].

Expt. 5. Dialysis of starch and salt solution. Make a 2 % solution of starch in water, as in Expt. 4 (*a*), and mix it with an equal volume of a 2 % solution of sodium chloride in water. Pour the mixture into a parchment dialyzer and dialyze in a beaker of distilled water. (The dialyzer should first be carefully tested to ascertain of there be a leak.) Test the liquid in the beaker with solutions of both iodine and silver nitrate. Some precipitate of chloride will be given, but no blue colour with iodine. After 24 hours, test the liquid again. There will be an increased amount of silver chloride formed, but a negative result with iodine. On addition of iodine to the liquid in the dialyzer, a blue colour is obtained. Hence we may assume that the colloidal starch does not pass through the membrane.

Some substances, such as gelatine (animal) and agar (vegetable), are only in the emulsoid condition at a raised temperature. When cold they set to form a " gel," in which the particles of the dispersed phase are no longer separate but united to make a solid. Silicic acid, the best known inorganic emulsoid, also sets to a gel on standing, either spontaneously or on addition of electrolytes. It is of classical interest since it was the substance largely used by Graham, the first worker on colloids.

Expt. 6. Preparation of gels. (*a*) *Agar*. Weigh out 2 gms. of agar and put it to soak in 100 c.c. of distilled water for an hour or two. Then boil: the agar gives a thick opalescent solution (sol) which sets to a gel on cooling. On warming, the gel again becomes a sol, and, on cooling, again sets to a gel. Thus the change is a reversible one. Agar is a mucilage which is obtained from certain genera of the Rhodophyceae (see p. 51). (*b*) *Silicic acid*. Weigh out 20 gms. of commercial "water-glass" syrup (a concentrated solution of sodium silicate) and dilute with 100 c.c. of freshly boiled distilled water (free from carbon dioxide). Pour 75 c.c. of this solution into a mixture of 25 c.c. of concentrated hydrochloric acid and 75 c.c. of water. Dialyze the mixture in a parchment dialyzer against running water for 3-4 hours. If to the dialyzed liquid a little very dilute ammonia is added, a gel will be formed in the course of a few hours. In this case, however, the process is irreversible, that is the gel cannot be reconverted again into the sol.

An interesting point in connexion with the colloidal state is that emphasized by Ostwald, i.e. that this condition is a *state*, not a *type*, of matter. Further, substances in the colloidal state do not constitute a definite class. It is reasonable to suppose that all substances which exist in the colloidal state can, under suitable conditions, also exist in the crystalline state, and *vice versa*. Further, the continuous phase is not always water. Sodium chloride, which is a very definite crystalloid, can be obtained in the colloidal state in petroleum ether. Most metals, even the alkali metals, have been obtained in colloidal solution: also a great many metallic oxides, hydroxides and sulphides.

The colloidal phases so far dealt with can be tabulated as follows[1]:

disperse		continuous	
liquid	......	solid	...... gels
solid	......	liquid	...... suspensoids
liquid	......	liquid	...... emulsoids

Some of the properties of colloidal solutions may now be considered. A point that has already been emphasized in the previous chapter is that the surface of particles in the colloidal state is very great in proportion to their mass. Such particles, moreover, unlike ions and small molecules in true solution, possess the properties of the surfaces of matter in mass, as, for instance, those connected with surface tension, electrical charge, etc., and these are especially marked on account of the proportionately large surfaces involved. Other properties are their inability, as a rule, to exert an osmotic pressure, to raise the boiling point, and to lower the freezing point of water. Some of the metallic suspensoids are characterized by their colour, this being red, purple or blue as in the case of gold sols.

An apparatus, by means of which the colloidal state can be demonstrated ocularly, is the ultramicroscope. This is a special form of microscope in which a powerful beam of light is directed upon a colloidal solution, which is then seen to contain a number of particles in rapid motion. When analyzed by special methods, this motion has been found to be identical with that shown by much larger, though still microscopic, particles, which has been termed Brownian movement.

Expt. 7. Demonstration of Brownian movement of microscopic particles. Mount a little gamboge in water and examine under the high power of a microscope. The particles will be seen to be in rapid motion.

It has been shown that Brownian movement is the outcome of the movement of the molecules of the liquid in which the particles are suspended. This movement is one of the factors which keeps the sol stable and prevents the particles from "settling" as in the case of a true suspension.

Another factor tending to keep the sol stable is the electrical charge borne by the particles. It is commonly known that there is usually a difference of potential between the contact surfaces of phases. If the

[1] There are also the following combinations (Bayliss, 1):

disperse		continuous	
gas	....	liquid	.... foam
liquid	....	gas	.... fog
solid	....	gas	.... tobacco smoke
solid	....	solid	.... ruby glass (colloidal sol of gold in glass).

particles in a colloidal solution all have the same charge, then they will tend to repulse one another mutually. It is found that the particles are charged, but the origin of the charge is not always clear. Sometimes if the substance in colloidal state is capable of electrolytic dissociation, the charge may arise in this way. Substances, however, as already mentioned, which are not dissociated may also bear a charge, and most frequently it is a negative one. It follows, then, that when an electrolyte is added to a colloidal solution, the charges on the colloidal particles are neutralized by the oppositely charged ions of the electrolyte, and they coalesce together and are precipitated.

As regards their behaviour to electrolytes the two classes, suspensoids and emulsoids, are very different. The suspensoids are very sensitive to traces of electrolytes, and, as they usually have a negative charge, it is the cation of the electrolyte which is the active ion; and of such, less of a divalent ion, than of a monovalent ion, is needed for precipitation and still less of a trivalent ion.

The emulsoids are far less sensitive to electrolytes in solution than the suspensoids; in fact, electrolytes, such as neutral alkali salts, must be added in very large quantities to emulsoids before precipitation takes place. Also, as a rule, whereas the precipitation of suspensoids is irreversible, that of emulsoids is reversible, that is, they pass into solution again on addition of water. In the case of an emulsoid in neutral solution this form of precipitation, unlike that of the suspensoids, may be regarded as consisting of two processes. First, a process analogous to that of "salting out" of soaps, esters, etc., in organic chemistry, which is, in effect, a withdrawal of water from one phase into another. Secondly, the precipitation is also affected to some extent by the valency of the ions of the salt used in precipitation.

When, however, a neutral solution of such an emulsoid as protein is made either slightly acid or alkaline, its behaviour towards neutral salts becomes altered. The precipitating power of salts in acid or alkaline medium is now in accordance with that on suspensoids. In alkaline solution the coagulating power of a salt depends on the valency of the cation; in an acid solution it depends on the valency of the anion.

The behaviour of proteins in acid and alkaline media is undoubtedly due to the fact that they are built up of amino-acids containing both amino and carboxyl groups. Such molecules may behave either as an acid or a base with the formation of salts. These are subject to electrolytic dissociation and hence acquire an electric charge. Such substances have been termed "amphoteric electrolytes" (see p. 134).

Expt. 8. *Precipitation of suspensoid sols by electrolytes.* The sols of gold, silver and arsenious sulphide carry an electro-negative charge: hence they are most readily precipitated by di- or tri-valent positive ions, such as Ba″ or Al‴. Add a few drops of barium chloride solution to the three sols (Expt. 3) respectively, and note that they are precipitated, though some time may elapse before the precipitation is complete. The ferric hydroxide sol, on the contrary, carries a positive charge. Hence it is most readily precipitated by a sulphate or phosphate. If a drop of sodium sulphate solution is added while the sol is hot, it is immediately precipitated.

Expt. 9. *Precipitation of emulsoid sols by electrolytes.* Saturate the starch, protein and soap solutions prepared in Expt. 4 with solid ammonium sulphate. Precipitation takes place in each case, and it is seen how large a quantity of electrolyte is needed for the "salting out" of emulsoid sols. Filter off the protein precipitate and suspend in distilled water. It will go into solution again, showing that the reaction is reversible.

REFERENCES

1. **Bayliss, W. M.** Principles of General Physiology. London, 1927. 4th ed.
2. **Burton, E. F.** The Physical Properties of Colloidal Solutions. London, 1921. 2nd ed.
3. **Hatschek, E.** An Introduction to the Physics and Chemistry of Colloids. London, 1925. 5th ed.
4. **Philip, J. C.** Physical Chemistry: its Bearing on Biology and Medicine. London, 1925. 3rd ed.
5. **Taylor, W. W.** The Chemistry of Colloids. London, 1921. 2nd ed.

CHAPTER III

PLANT ENZYMES

SOME indication has been given in the previous chapter of the large number of complex processes which take place in the plant, and it has been mentioned that many of these are controlled by enzymes.

The most remarkable feature in connexion with the chemical processes of plant metabolism is the ease and rapidity with which, at ordinary temperatures, chemical reactions take place, when under artificial conditions they need a much longer time and higher temperatures.

It has been found that many of the chemical reactions in the plant can be brought about *in vitro* on addition of certain substances which can be extracted from the plant. These substances are known as enzymes. It is the property of enzymes that they are able to accelerate reactions which, in their absence, would only take place very slowly. The enzyme cannot initiate these reactions and does not form part of their final products.

Some inorganic substances have the same property of accelerating reactions, and such substances are termed catalysts. For example, when water is added to ethyl acetate, the latter begins to be decomposed slowly into ethyl alcohol and acetic acid:

$$\text{ethyl acetate} + \text{water} \longrightarrow \text{ethyl alcohol} + \text{acetic acid},$$

but if, in addition, some hydrochloric acid is added, hydrolysis takes place with much greater rapidity, and at the end of the reaction the hydrochloric acid is found unchanged. Hence in this case hydrochloric acid is an inorganic catalyst. Many other similar instances are known as, for example, the catalyzing effect of a small quantity of manganese dioxide which brings about the liberation of oxygen from potassium chlorate at a much lower temperature than by heat alone.

By analogy, therefore, an enzyme may be defined as an organic catalyst produced by the plant.

Another point in connexion with the above-mentioned reaction of water with ethyl acetate, is the fact of its being representative of the type known as reversible. After a certain amount of acetic acid and ethyl alcohol has been formed, these recombine to form ethyl acetate until in time a certain point of equilibrium is reached. Since the same point of equilibrium is reached whether hydrochloric acid is used or not, it is obvious that the hydrochloric acid accelerates the reaction in both directions:

$$\text{ethyl acetate} + \text{water} \rightleftharpoons \text{ethyl alcohol} + \text{acetic acid.}$$

Such a reaction is termed a reversible one. Many of the processes accelerated by enzymes in the plant are reversible, and there is reason to believe that the enzyme accelerates the reaction in both directions.

The substance upon which the enzyme acts is termed the substrate, and it is supposed that some kind of loose combination occurs between these two substances. The enzyme is unaltered when the reaction is complete, unless it is affected by the products formed.

The enzymes are very widely distributed and form constituents of all living cells, though all tissues do not necessarily contain the same enzymes.

There is no doubt that many enzymes are specific, in which case an enzyme can only accelerate one reaction, or one class of reaction. We cannot be sure that any enzyme is specific and different from all others, until it has been proved that it accelerates one process which is incapable of being accelerated by any other enzyme. It is possible that some enzymes, to which separate names have been given, are really identical.

Most of the plant enzymes are soluble in water and dilute glycerol and sometimes in dilute alcohol. Some can be extracted by simply macerating the tissues with water; others are more intimately connected with the protoplasm, and are only extracted if the protoplasm is killed by certain reagents, of which those most frequently employed are toluol and chloroform. These substances kill the protoplasm and do not, in many cases, affect the enzyme. After the death of the protoplasm, the enzymes are more readily extracted from the cell. From aqueous solutions enzymes can usually be precipitated by adding strong alcohol.

The majority of enzymes are destroyed by raising the temperature above 60° C. *In vitro* their reactions are generally carried out most rapidly between the temperatures of 35–45° C.

In performing experiments with enzymes *in vitro*, it is always necessary to add an antiseptic, otherwise the reaction to be studied will be masked or entirely superseded by the action of bacteria unavoidably present. Toluol and chloroform mentioned above, as well as thymol, may be used. These reagents prevent any bacterial action from taking place. Some enzymes, however, are susceptible to chloroform, as, for instance, maltase.

The chemical nature of enzymes is at present unknown, because it is difficult to purify them without destroying them, and hence to obtain them of sufficient purity for chemical analysis. They were originally thought to be proteins, but with the improvements in methods for purification, it has been found that the protein reactions disappear, although the enzyme activity does not decrease. In solution they exist in the colloidal condition.

The questions as to their origin and their relation to the protoplasm cannot yet be answered with any certainty. It is also impossible to say whether the majority of chemical processes in the plant are catalyzed by enzymes.

A feature of enzyme action which is of considerable interest and which has already been mentioned is the question as to whether enzymes catalyze a reaction in both directions. Thus, in the case of hydrolytic enzymes which constitute by far the greater number of known enzymes, do they control the synthetic as well as the hydrolytic process? There is evidence that this is so, since, in many cases, the hydrolysis is not complete. If the enzyme were a catalyst in one direction only, the reaction would be complete. Further evidence is supplied by the fact that, under suitable conditions, i.e. strong concentration of the substances from which synthesis is to take place, certain syntheses have been carried out *in vitro*. As an example may be quoted the synthesis of maltose from a concentrated solution of glucose by maltase (Bayliss, 2).

In the living cell it is supposed that the hydrolysis and synthesis are balanced. On the "death" of the protoplasm, which may be caused by mechanical injury, vapour of chloroform or toluol, etc. (Armstrong, 7, 8), the reactions catalyzed by enzymes cease to be balanced and proceed almost always in the direction of hydrolysis and the splitting up of more complex into simpler substances. This phenomenon is obvious when any of the products can be recognized by smell or colour, as, for instance, the smell of benzaldehyde on injuring leaves of plants containing cyanogenetic glucosides (see p. 161), or the production of coloured oxidation products when some of the aromatic glucosides are decomposed (see p. 124).

If plant tissues are disintegrated, and the mass is kept at a temperature of about 38° C., the above-mentioned hydrolytic processes continue to be catalyzed by the enzymes present until equilibrium is reached, which will be near complete hydrolysis, especially if water has been added. Such a process is termed "autolysis."

The chief plant enzymes may be classified to a certain extent according to the reaction they catalyze, e.g. hydrolytic, oxidizing, etc., as follows:

Hydrolysis

Enzyme	Substrate	Products
Lipase (p. 94)	Fats	Fatty acids and glycerol
„ (p. 99)	Lecithin	Fatty acids, glycero-phosphoric acid and choline
Chlorophyllase (p. 34)	Chlorophyll	Chlorophyllide and phytol
Phytase (p. 102)	Phytin	Inositol and phosphoric acid
Glycerophosphatase (p. 99)	Glycerophosphoric acid	Glycerol and phosphoric acid
Diastase (p. 75)	Starch	Dextrin and maltose
Invertase (p. 78)	Cane sugar	Dextrose and laevulose
Maltase (p. 77)	Maltose	Dextrose
Inulase (p. 60)	Inulin	Laevulose
Cytase (p. 71)	Hemicellulose	Mannose and galactose
Emulsin (p. 160)	Amygdalin	Benzaldehyde, prussic acid and glucose
Myrosin (p. 164)	Sinigrin	Allyl isothiocyanate, potassium, hydrogen sulphate and glucose
Pepsin (p. 152)	Proteins	Albuminoses and peptones
Erepsin (p. 152)	Peptones	Polypeptides and amino-acids

Oxidation and reduction

Peroxidase (p. 122)	Hydrogen peroxide	Atomic oxygen
Oxygenase (p. 122)	Catechol, etc	Peroxide
Tyrosinase (p. 128)	Tyrosine	Melanin
Catalase (p. 129)	Hydrogen peroxide	Molecular oxygen
Reductase (oxido-reductase) (p. 129)	Water	Hydrogen and oxygen

Respiration (and fermentation)

Hexosephosphatase (p. 22)	Hexosephosphate	Hexose and phosphoric acid
Zymase (p. 22)	Hexose	Alcohol and carbon dioxide
Carboxylase (p. 22)	Pyruvic acid, etc.	Acetaldehyde and carbon dioxide

Other reactions

Urease (p. 181)	Urea	Ammonia and carbon dioxide
Pectase (p. 67)	Soluble pectin	Pectic acid

Most of these various classes of enzymes will be dealt with in detail in connexion with the chemical substances on which they react.

An excellent demonstration of the fact that a single cell may contain all the various enzymes connected with the processes of metabolism is afforded by the unicellular Fungus, Yeast (*Saccharomyces*), of which many species and varieties are known. The feature of special interest in connexion with the Yeast plant is its power of fermenting hexoses, with the formation of alcohol and carbon dioxide, the process being carried out by means of an enzyme, zymase. The complete reaction is generally represented as follows:

$$C_6H_{12}O_6 = 2CO_2 + 2C_2H_5OH$$

though there is little doubt that several stages are involved, including oxidation, reduction and hydrolysis. It has been known for some time that phosphates are essential to the action of zymase, and the first stage is probably the formation of a hexosephosphate with the accompanying production of ethyl alcohol and carbon dioxide:

$$2C_6H_{12}O_6 + 2R_2HPO_4 = C_6H_{10}O_4(R_2PO_4)_2 + 2C_2H_5OH + 2CO_2 + 2H_2O,$$

the hexosephosphate being continually decomposed by a hydrolytic enzyme, hexosephosphatase, yielding free phosphate again:

$$C_6H_{10}O_4(R_2PO_4)_2 + 2H_2O = C_6H_{12}O_6 + 2R_2HPO_4.$$

In addition to zymase and hexosephosphatase, yeast contains the enzymes, invertase, protease, peroxidase, catalase, reductase, glycogenase, carboxylase, a glucoside-splitting enzyme, and some form of diastatic enzyme. The carboxylase decomposes a large number of aliphatic α-keto-acids, of which the most important is pyruvic acid $CH_3 \cdot CO \cdot COOH$. The reaction, which is also possibly concerned in fermentation, involves the formation of the corresponding aldehyde with the evolution of carbon dioxide:

$$CH_3 \cdot CO \cdot COOH = CH_3 \cdot CHO + CO_2.$$

Yeast also stores, as a reserve material, the polysaccharide, glycogen, which occurs in animal tissues though it is rarely found in plants: this is hydrolyzed by glycogenase into a monosaccharide. Finally, yeast contains invertase, and most species, in addition, maltase, but from a few species the latter enzyme is absent. Hence yeasts are able to ferment the disaccharides, cane-sugar and maltose, since they can first hydrolyze them to monosaccharides.

As in the case of the enzymes of other tissues, those of yeast can be made to carry out their functions after the death of the living protoplasm. The method of demonstrating this is to "kill" the cells by means of drying at 25–30° C., by treatment with a mixture of alcohol and ether, or by treatment with acetone and ether. In this way the protoplasm is destroyed, but the enzymes remain uninjured. Yeast treated thus has been termed "zymin."

From zymin some of the enzymes, e.g. invertase and the glucoside-splitting enzyme, can be extracted with water: other enzymes, e.g. zymase and maltase, are not so readily extracted. From the living cells the enzymes are only obtained with difficulty, the extraction of yeast juice, containing zymase and other enzymes, needing, by Buchner's method, a pressure as great as 500 atmospheres.

In connexion with alcoholic fermentation by zymase, the following point is of special interest. For carrying out this process, another substance is necessary in addition to the phosphate and enzymes already mentioned, i.e. a thermostable co-enzyme of unknown nature. The separation of zymase from the co-enzyme has been accomplished by filtering expressed (Buchner) yeast juice through a special form of gelatine filter under a pressure of 50 atmospheres. The phosphate and co-enzyme can also be removed from zymin by washing with water. The washed residue is then found to be incapable of fermentation, as also are the washings. If, however, the boiled washings are added to the washed residue, the system is synthesized and can now carry out fermentation again. The chemical nature of the co-enzyme, which is thermostable, and the precise part played by it in the process, are as yet unknown (Harden, 4).

Expt. 10. *Preparation of zymin.* Take 50 gms. of baker's yeast and stir it into 300 c.c. of acetone. Continue stirring for 10 minutes, and filter on a filter-pump. The mass is then mixed with 100 c.c. of acetone for 2 minutes and again filtered. The residue is roughly powdered, well-kneaded with 25 c.c. of ether for 3 minutes, filtered, drained and spread on filter-paper for an hour in the air. It can be finally dried at 45° C. for 24 hours.

Expt. 11. *Action of zymase.* (a) *Detection of carbon dioxide.* It has been shown (Harden, 4) that the greater the volume of sugar solution used with a given weight of zymin, the weaker is its action. To demonstrate its activity, therefore, it is best to use not more than 5–10 c.c. of 10 % glucose solution for every 2 gms. of zymin. Place the mixture in a test-tube and fit it with a cork and glass tubing, the latter dipping under a solution of lime water in a test-tube. Place the test-tube containing the zymin and glucose solution in a beaker of water and warm to 35–40° C. Bubbles of carbon dioxide will be evolved and will produce a precipitate of calcium carbonate in the lime water. A control experiment should be made using boiled zymin. (b) *Detection of alcohol.* Into a small flask put 8 gms. of zymin, 20 c.c. of 10 % glucose solution and a little toluol. Keep the flask in an incubator at 37–40° C. for 12 hours. Then filter through filter-paper (or linen) into a small distilling flask. Distil over one half or two-thirds of the original volume. Add to the distillate in a test-tube, 3–5 c.c. of iodine in potassium iodide solution and then 5 % caustic soda until the colour vanishes. Shake up and warm gently in a beaker of water to 60° C. A smell of iodoform will be detected and a yellow crystalline deposit of the same substance will appear in the tube on cooling and standing. Examine the crystals under the microscope and note their characteristic star-like shape.

Expt. 12. *Action of maltase.* (Harden and Zilva, 12.) Into each of two small flasks, put 20 c.c. of a 2 % solution of maltose and 0·5 gm. of zymin. Boil the contents of one flask. Then plug both flasks with cotton-wool, add a few drops of toluol and place in an incubator at 38° C. for 12–24 hours. Filter the liquid from both flasks and test by making the osazone (see p. 50), using at least 10 c.c. of the filtrate in each case. Glucosazone will crystallize out from the unboiled, maltosazone from the boiled, mixture.

Expt. 13. *Action of carboxylase.* (Harden, 10.) The action of carboxylase on pyruvic acid is detected by the formation of carbon dioxide and acetaldehyde. Carefully prepared zymin will still respire, but, after washing, some constituent essential to respiration is removed. Hence the zymin must be first washed and tested. Take 5 gms. of zymin and wash well on a filter with distilled water. Then suspend the zymin in 50 c.c. of water in a flask and draw a slow current of air (previously passed through two bottles of strong caustic soda and two bottles of saturated baryta solution) through the suspension of zymin into a receiving flask of baryta solution. The flasks should be connected with pressure tubing and the apparatus must be air tight. Continue to draw the current of air through until it ceases to produce a milkiness in the receiving flask, due to any carbon dioxide in solution or to residual respiration. Then add quickly to the suspension of zymin 50 c.c. of 1 % pyruvic acid (by weight), 5 c.c. of normal caustic potash and 6 gms. of boric acid; also a few drops of caprylic alcohol to prevent frothing. Place the flask in a beaker of water at 30–40° C. and again draw a current of air. A copious precipitate of barium carbonate will be formed in the receiving flask. The boric acid is used to prevent the solution from becoming too alkaline owing to the formation of potassium carbonate, and, being a weak acid, it has no inhibiting action on the enzyme.

The contents of the flask containing the zymin are filtered into a small distilling flask and about 10 c.c. of distillate collected (cooled with ice if possible). To this add 1–2 c.c. of a freshly made 1 % solution of sodium nitroprusside, followed by a few drops of piperidine. A deep blue colour denotes the presence of acetaldehyde.

Expt. 14. *Action of peroxidase* (Harden and Zilva, 12.) Into four small evaporating dishes, (*a*), (*b*), (*c*) and (*d*), put the following :

(*a*) A suspension of 0·5 gm. of fresh yeast in 10 c.c. distilled water + 1 c.c. of benzidine solution (1 % in 50 % alcohol) + 2–3 drops of hydrogen peroxide (20 vols.).

(*b*) A suspension of 0·5 gm. of zymin in 10 c.c. of distilled water + 1 c.c. of benzidine solution + 2–3 drops of hydrogen peroxide.

(*c*) A suspension of 0·5 gm. of washed zymin in 10 c.c. of distilled water + 1 c.c. of benzidine solution + 2–3 drops of hydrogen peroxide. (The zymin is washed by putting it on a double folded filter-paper in a funnel and adding distilled water from time to time. 50 c.c. of water should be used for 0·5 gm. of zymin.)

(*d*) A suspension of 0·5 gm. of washed zymin in 10 c.c. of washings + 1 c.c. of benzidine solution + 2–3 drops of hydrogen peroxide.

A blue colour will develop in (*a*) showing that fresh yeast contains a peroxidase (see p. 124). A blue colour will also develop in (*c*) but not in (*b*) and (*d*). This is explained by assuming that the zymin contains an inhibitor, not present in fresh yeast, but which is developed during the preparation of the zymin, and that this inhibitor can be washed away by water. On adding the washings to the washed zymin the reaction is inhibited again.

Expt. 15. *Action of catalase.* (Harden and Zilva, 12.) Completely fill a test-tube with hydrogen peroxide (20 vols.) solution which has been diluted with an equal volume of water and add 0·5–1 gm. of zymin. Place the thumb firmly over the mouth of the tube, invert and place the mouth under water in a small basin, clamping the tube in position. A rapid evolution of oxygen takes place. When the tube is about three-fourths full of gas, close the mouth with the thumb while still under water and remove the tube. Plunge a glowing splint into the gas and it will re-kindle to a flame.

Expt. 16. *Action of protease.* Weigh out 10 gms. of white flour, and allow it to extract with 100 c.c. of distilled water for one hour, shaking from time to time. Then filter on a filter-pump. The extract will contain the albumin, leucosin (see p. 138). Into small flasks (*a*) and (*b*) put the following:

(*a*) 40 c.c. of the flour extract + 1 gm. of zymin + 1 c.c. of toluol.

(*b*) 40 c.c. of flour extract + 1 gm. of boiled zymin + 1 c.c. of toluol.

Shake both flasks, plug with cotton-wool and place them in an incubator at 38° C. for 48 hrs. After incubation, boil the liquid in both flasks, in order to coagulate un-altered protein, and filter. Cool the filtrates from the respective flasks and add bromine water drop by drop (see p. 153). A pink, or purplish-pink colour, due to the presence of tryptophane, will be formed in tube (*a*). Hence hydrolysis of protein has taken place. Tube (*b*) will show no colour or only that due to bromine. Add a *little* amyl alcohol to both tubes and shake gently. The alcohol will be coloured pink or purplish in the tube giving the tryptophane reaction.

Expt. 17. *Action of reductase.* (Harden and Norris, 11.) Take two test-tubes, (*a*) and (*b*), provided with well-fitting corks and put in the following:

(*a*) 1 gm. of zymin + 20 c.c. of distilled water + 0·5 c.c. of methylene blue solu-tion (made by diluting 5 c.c. of a saturated alcoholic solution to 200 c.c. with distilled water).

(*b*) 1 gm. of boiled zymin + 20 c.c. of distilled water + 0·5 c.c. of methylene blue solution.

Cork both tubes after adding a few drops of toluol and place in an incubator at 38° C. for 1–3 hours. The blue colour will practically disappear from tube (*a*) but will remain in tube (*b*).

The methylene blue is reduced to a colourless leuco-compound which will become blue again on re-oxidation.

Expt. 18. *Enzyme actions of an aqueous extract of zymin.* Weigh out 2 gms. of zymin and place them on a double folded filter-paper in a funnel and wash with 80 c.c. of distilled water. With the filtrate make the following experiments.

(A) *Action of invertase.* (Harden and Zilva, 12.) Into two small flasks (*a*) and (*b*) put the following:

(*a*) 10 c.c. of a 2% solution of pure cane-sugar + 10 c.c. of the filtrate from zymin.

(*b*) 10 c.c. of the same solution of cane-sugar + 10 c.c. of the boiled filtrate from zymin.

Put both flasks in an incubator at 38° C. After 30 minutes add equal quantities (about 1–2 c.c.) of Fehling's solution to both flasks and boil (see p. 54). Flask (*a*) will show considerable reduction of the Fehling. Flask (*b*) will show comparatively little reduction, that which does take place probably being due to the sugar previously formed by the action of glycogenase on stored glycogen.

(B) *Action of the glucoside-splitting enzyme.* (Caldwell and Courtauld, 9; Henry and Auld, 13.) This enzyme will act upon the glucoside, amygdalin, which is present in bitter almonds, with the production of glucose, benzaldehyde and prussic acid (see p. 160). Into two small flasks (*a*) and (*b*) put the following:

(*a*) 20 c.c. of a 2 % solution of amygdalin + 20 c.c. of the filtrate from zymin.

(*b*) 20 c.c. of the same solution of amygdalin + 20 c.c. of the boiled filtrate from zymin.

Add a few drops of toluol to both flasks and then cork, inserting, with the cork, a strip of paper which has been dipped in solutions of picric acid and sodium carbonate (see p. 161). Put both flasks in an incubator at 38° C. for 12–24 hours. The picrate paper in flask (*a*) will have reddened. Add a little Fehling's solution to the liquid in the same flask and boil. The Fehling will be reduced. The liquid in flask (*b*) will only reduce Fehling slightly [see Expt. A (*b*)] and the picrate paper will not be reddened.

REFERENCES

Books

1. **Abderhalden, E.** Biochemisches Handlexikon, v. Berlin, 1911.
2. **Bayliss, W. M.** The Nature of Enzyme Action. London, 1925. 5th ed.
3. **Euler, H.** Chemie der Enzyme. München und Wiesbaden, 1920–27.
4. **Harden, A.** Alcoholic Fermentation. London, 1923. 3rd ed.
5. **Vernon, H. M.** Intracellular Enzymes. London, 1908.
6. **Waldschmidt-Leitz, E.** Die Enzyme. Braunschweig, 1926.
7. **Wohlgemuth, J.** Grundriss der Fermentmethoden. Berlin, 1913.

Papers

7. **Armstrong, H. E.**, and **Armstrong, E. F.** The Origin of Osmotic Effects. III. The Function of Hormones in Stimulating Enzymic Change in Relation to Narcosis and the Phenomena of Degenerative and Regenerative Change in Living Structures. *Proc. R. Soc.*, 1910, B Vol. 82, pp. 588–602. *Ibid.* IV. Note on the Differential Septa in Plants with reference to the Translocation of Nutritive Materials. *Proc. R. Soc.*, 1912, B Vol. 84, pp. 226–229.

8. **Armstrong, H. E.**, and **Armstrong, E. F.** The Function of Hormones in regulating Metabolism. *Ann. Bot.*, 1911, Vol. 25, pp. 507–519.

9. **Caldwell, R. J.**, and **Courtauld, S. L.** Studies on Enzyme Action. IX. The Enzymes of Yeast: Amygdalase. *Proc. R. Soc.*, 1907, B Vol. 79 pp. 350–359.

10. **Harden, A.** The Enzymes of Washed Zymin and Dried Yeast. I. Carboxylase. *Biochem. J.*, 1913, Vol. 7, pp. 214–217.

11. **Harden, A.**, and **Norris, R. V.** The Enzymes of Washed Zymin and Dried Yeast. II. Reductase. *Biochem. J.*, 1914, Vol. 8, pp. 100–106.

12. **Harden, A.**, and **Zilva, S. S.** The Enzymes of Washed Zymin and Dried Yeast. III. Peroxydase, Catalase, Invertase and Maltase. *Biochem. J.*, 1914, Vol. 8, pp. 217–226.

13. **Henry, T. A.**, and **Auld, S. J. M.** On the Probable Existence of Emulsin in Yeast. *Proc. R. Soc.*, 1905, B Vol. 76, pp. 568–580.

CHAPTER IV

CHLOROPHYLL

THE fact has already been emphasized that the plant synthesizes all the complex organic substances of which it is built from the simple compounds, carbon dioxide, water and inorganic salts. The initial metabolic process and the one from which all others have their starting-point is that of a synthesis of a carbohydrate from carbon dioxide and water. This synthesis can only be carried out in the light, and only in a green plant, i.e. a plant containing chlorophyll. Chlorophyll may almost be considered the chemical substance of primary importance in the organic world, for upon it depends the life of all plants and animals. Animals depend for their existence on certain complex amino-acids, some of which they are unable to synthesize for themselves, and which they derive from plants. Plants in turn are unable to exist except by virtue of the properties of chlorophyll.

The property of chlorophyll which is so important is the power it possesses of absorbing the radiant energy of the sun's rays and converting it into chemical energy by means of which a carbohydrate is synthesized. This summarizes the whole process, which, however, can scarcely be very simple, and probably consists of several reactions at present undifferentiated. If the formula for carbonic acid is compared with that of a simple carbohydrate such as a tetrose, pentose or hexose, the following relationship is seen:

$$H_2CO_3 \rightarrow (H_2CO)_x \text{ where } x = 4, 5 \text{ or } 6,$$

that is, in the synthesis of a carbohydrate a reducing reaction must take place.

Many hypotheses have been formulated as to the nature of these reactions. The one which has most frequently been advanced suggests that formaldehyde is the first product of the synthesis from carbon dioxide and water which takes place in the green plant; that the reaction involves reduction with elimination of oxygen:

$$H_2CO_3 = H_2CO + O_2,$$

and that this product is later condensed to form a hexose,

$$6H_2CO = C_6H_{12}O_6.$$

As the concentration of sugar increases in the cell, further condensation may take place to form starch :

$$x\,(C_6H_{12}O_6) = (C_6H_{10}O_5)_x + x\,H_2O.$$

The facts in agreement with these views are: first, in most plants a volume of oxygen is given off approximately equivalent to the volume of carbon dioxide absorbed; secondly, in some plants starch, in others sugar, is known to be produced during photosynthesis. The detection of formaldehyde, either in the plant or in certain systems containing chlorophyll, as a proof of its formation during photosynthesis, has been shown to be invalid (see p. 37) (Jörgensen and Kidd, 2).

The value and significance of this reducing reaction is seen when it is realized that, by oxidation of the carbohydrates synthesized, energy is produced to supply the needs of the whole metabolism of the plant (see p. 6).

In the chemical treatment of the subject of carbon assimilation, some of the chemical properties of chlorophyll will first be considered, and, later, its behaviour under certain conditons: the chemistry, however, of the phenomenon itself is as yet unknown.

The following account, as far as it concerns chlorophyll, and the accompanying experiments are taken from a *résumé* (Jörgensen and Stiles, 3) of the original work (Willstätter und Stoll, 1) upon which the entire knowledge of the subject is based.

CHLOROPHYLL.

Our knowledge of the chemistry of chlorophyll has, within recent years, been set upon a firm experimental basis (Willstätter und Stoll, 1). The results which have been arrived at may broadly be summarized as follows:

In all plants examined the chloroplastids contain four pigments, of which two (termed respectively chlorophylls a and b) are green, and two are yellow. They occur in about the following proportions in fresh leaves:

Green $\begin{cases} \text{Chlorophyll } a \ \ldots \ C_{55}H_{72}O_5N_4Mg \ \ldots \ 2 \text{ pts per } 1000 \\ \text{Chlorophyll } b \ \ldots \ C_{55}H_{70}O_6N_4Mg \ \ldots \ \frac{3}{4} \quad ,, \qquad ,, \end{cases}$

Yellow $\begin{cases} \text{Carotin} \ \ \ldots\ldots\ldots \ C_{40}H_{56} \ \ldots\ldots\ldots\ldots \ \frac{1}{6} \quad ,, \qquad ,, \\ \text{Xanthophyll} \ \ldots \ C_{40}H_{56}O_2 \ \ldots\ldots\ldots \ \frac{1}{3} \quad ,, \qquad ,, \end{cases}$

A point of great interest in connexion with chlorophyll is that it contains magnesium to the extent of $2 \cdot 7\,\%$ but no other metal is present. **Chlorophyll a,** when isolated, is a blue-black solid giving a green-blue solution in the solvents in which it is soluble, i.e. ethyl alcohol, acetone, chloroform, ether, carbon bisulphide, pyridine and benzene. **Chlorophyll b**, when isolated, is a green-black solid giving a pure green solution: it has much the same solubilities as chlorophyll *a*. The two chlorophylls, however, can be separated by their different solubilities in methyl alcohol. Both can be obtained in microscopic crystals.

Carotin crystallizes in orange-red crystals, and xanthophyll in yellow crystals.

In the chloroplastids these pigments occur mixed with various colourless substances, fats, waxes, and salts of fatty acids.

When chlorophyll is spoken of, it will be understood to refer to the green pigments and not to the yellow.

The *pure* pigments, when *isolated*, are readily soluble in acetone, ether and benzene. When very thoroughly dried nettle leaves are treated with pure acetone, no green colour is extracted, but if a few drops of water are added, the extract becomes green. Also if acetone is poured on to fresh leaves, the pigment is extracted. The explanation offered for these phenomena is that chlorophyll is present in a colloidal condition in the cell. This point will be considered again later (see p. 36).

The Common Nettle (*Urtica*) is the plant which has been used for material for the extraction of chlorophyll on a large scale, and it also forms very useful material for extraction on a small scale. The pigment has been found to be unaltered by drying, and, since dried leaves involve far less bulk and dilution of solvents, material should be dried before using. Some leaves (Elder and Conifers) are spoilt by drying. From dried leaves pure solvents, such as petrol ether, benzene and acetone, extract very little pigment for reasons which will be mentioned later, but if the solvents contain a moderate amount of water, the pigment is readily soluble. About 80 % acetone is the best solvent. The nettle leaves are removed from the stalks and laid on sheets of paper to dry. When well air-dried they are finely powdered, and the powder further dried at 30-40° C. in an incubator. The leaf-powder can be kept for a considerable time in a well-stoppered bottle.

Expt. 19. *Extraction of pigment.* Two grams of leaf-powder are sucked to a filter-paper on a small porcelain funnel and 2-3 c.c. of 85 % acetone are added. This is allowed to soak into the powder for a few minutes. The fluid is then sucked through with the pump, the flask disconnected and more acetone added. The operation is repeated until 20 c.c. of the solvent have been added, when the powder is sucked dry. A deep blue-green solution with a red fluorescence is obtained which contains all the four pigments from the leaf. The acetone extract thus obtained is then poured into double the quantity of *petrol ether* contained in a separating funnel. An equal quantity of distilled water is added, this being poured gently down the side of the funnel in order to avoid the formation of emulsions. In the course of a few minutes, the ether layer separates out and now contains the pigments. The lower layer, which is slightly green, is run off. The addition of distilled water and subsequent removal of the lower layer is repeated about four times, in order completely to remove the acetone from the ether solution. If the ether solution should have become at all emulsified, it can be cleared by shaking with anhydrous sodium sulphate and filtering.

The whole process should be repeated with another 2 gms. of leaf-powder and the pigment transferred to *ether*, since a solution in this solvent is required for later experiments.

Expt. 20. *Demonstration of the presence of chlorophylls a and b.* Of the petrol ether solution from the last experiment, 10 c.c. are shaken with 10 c.c. of 92% methyl alcohol. Two layers are formed of which the petrol ether layer contains chlorophyll *a*, and the methyl alcohol layer chlorophyll *b*. The solution of chlorophyll *a* is blue-green, while that of chlorophyll *b* is a purer green, but the colour difference between them is diminished owing to the presence of the yellow pigments, of which carotin is in the petrol ether, and xanthophyll in the methyl alcohol. Keep the two extracts for Expt. 24.

As will be explained later, the green pigments of chlorophyll can be saponified by alkalies and are then insoluble in ethereal solution. This method can be adopted to separate the green from the yellow pigments, xanthophyll and carotin.

Expt. 21. *Separation of green and yellow pigments.* Shake 5 c.c. of an ether solution of the pigments (Expt. 19) with 2 c.c. of 30% caustic potash in methyl alcohol (obtained by dissolving 30 gms. of potassium hydroxide in 100 c.c. of methyl alcohol[1]). After the green colour has reappeared, slowly add 10 c.c. of water and then add a little more ether. On shaking the test-tube, two layers are produced, of which the lower watery-alkaline one contains the saponified green pigments, while the carotin and xanthophyll are contained in the upper ethereal layer.

Expt. 22. *Separation of the two yellow pigments.* The ether layer obtained in the last experiment is washed with water in a separating funnel, and evaporated down to 1 c.c. It is then diluted with 10 c.c. of petrol ether and next mixed with 10 c.c. 90% methyl alcohol. The methyl alcoholic layer is removed and the petrol ether layer is again treated with methyl alcohol and the methyl alcohol again removed. This process is repeated until the methyl alcohol is no longer coloured. The methyl alcohol contains the xanthophyll, the petrol ether the carotin.

Further accounts of the yellow pigments are given on p. 40.

The best known reactions of chlorophyll are those which take place with acids and alkalies respectively.

Chlorophyll is a neutral substance, and, on treatment with alkalies, it forms salts of acids, the latter being known as chlorophyllins. These salts are soluble in water forming green solutions which are not however

---

[1] The methyl alcohol must be very pure, otherwise the alcoholic potash solution will become brown and discoloured.

fluorescent. Chlorophyll a may be represented as the methyl phytyl ester of an acid chlorophyllin (phytol is a primary alcohol, see p. 39):

$$C_{32}H_{30}ON_4Mg \Big\langle \begin{matrix} COOCH_3 \\ COOC_{20}H_{39} \end{matrix} \qquad C_{32}H_{30}ON_4Mg \Big\langle \begin{matrix} COOH \\ COOH \end{matrix}$$

Chlorophyll a Chlorophyllin

On treatment in the cold with alkali, the ester is saponified, and the alkali salt of chlorophyllin is formed. During saponification, there is a change of colour in the pigment, the so-called brown phase, followed by a return to green.

Expt. 23. Saponification of a mixture of the green pigments. Pour a little of the ether solution obtained in Expt. 19 into a test-tube, and in a pipette take a little 30 % solution of potash in methyl alcohol. Place the lower end of the pipette at the bottom of the test-tube and allow the potash to run in below the chlorophyll solution. At the interface between the solutions there appears immediately a brown-coloured layer which diffuses on shaking. In about ten minutes it changes back through an olive-green colour to pure green.

The chlorophyll has been saponified to the potassium salt of the acid chlorophyllin. This salt is insoluble in ether, so if water is added to bring about a separation of the two layers, the green colour is no longer present in the ethereal layer.

The change of colour on saponification is different for the two chlorophylls, the brown phase produced in the above mixture of chlorophylls being due to a yellow phase produced by chlorophyll a, and a brown-red phase produced by chlorophyll b. To demonstrate this the phase test (Expt. 23) may also be carried out separately on the two chlorophylls.

Expt. 24. Saponification of chlorophylls a and b separately. The methyl alcohol solution obtained in Expt. 20 is transferred to ether as in Expt. 19. Both the latter and the petrol ether solution of chlorophyll a are saponified as in the previous experiment.

As already demonstrated the potassium salts of the chlorophyllins which are produced by saponification of the mixture of green pigments in the cold are not fluorescent. By saponification of chlorophyll with hot alkali, isochlorophyllins are formed (see Expt. 25 below) which are fluorescent.

On *heating* chlorophyllins with concentrated alcoholic alkalies, a series of decomposition products, phyllins (also acids) are obtained by removal of carboxyl groups. The final phyllin has only one carboxyl group. When this is removed, a substance, aetiophyllin, $C_{31}H_{34}N_4Mg$, is obtained which contains no oxygen (see Scheme 1, p. 35).

Another difference between the results of treating chlorophyll with hot and cold alkali is that in the former process the yellow pigments are destroyed. If then water is added after saponification with hot alkali, and the solution is shaken up with ether, the ether will remain colourless.

When chlorophyll is treated with acids, a different reaction takes place. The chlorophyll changes in colour to olive-green and loses most of its fluorescence. The magnesium of the molecule is removed, being replaced by hydrogen, and the resulting product is termed phaeophytin (see Scheme 1, p. 35).

From phaeophytin a series of decomposition products have been obtained, which fall into two groups, the phytochlorins and the phyto-rhodins. The phytochlorins are olive-green in colour, and are derived from chlorophyll *a*; the phytorhodins are red, and are derived from chlorophyll *b*. The phaeophytins from the two chlorophylls are indistinguishable until the above decomposition products are obtained. (The original discovery of two kinds of chlorophyll was brought about by the differentiation of these decomposition products.)

A number of phytochlorins and phytorhodins have been identified and are designated by letters *a*, *b*, etc. By employing a uniform method of treatment, however, two of these products, phytochlorin *e* and phyto-rhodin *g*, can be secured.

The phytochlorins and the phytorhodins are of course magnesium-free compounds and can be obtained by the action of acid on the chlorophyllins and isochlorophyllins. Phytochlorin *e* and phytorhodin *g*, in particular are obtained by the action of acid on isochlorophyllins, i.e. they are magnesium-free isochlorophyllins. They are formed by the addition of acid to the products of saponification with hot alkali.

The separation of the various phytochlorins and phytorhodins can be brought about by means of their different distribution between ether and hydrochloric acid: each compound can be extracted from ether according to the concentration of the acid used.

Expt. 25. *The formation of phytochlorin and phytorhodin.* 5 c.c. of an ether solution containing both chlorophylls *a* and *b* are evaporated to dryness in a test-tube in a water-bath, and the residue treated with 3 c.c. of boiling 30 % potash solution in methyl alcohol, and boiled gently for half a minute. A liquid with a red fluorescence is produced which consists of a solution of the potassium salts of isochlorophyllins. The solution is diluted with double its volume of water, and concentrated hydrochloric acid is added until the solution is just acid. The liquid is then shaken with ether in a separating funnel: the dissociation products produced by the previous treatment pass into the ether solution which thus acquires an olive-brown colour.

The ether solution is shaken twice, each time with 10 c.c. of 4 % hydrochloric acid (sp. gr. 1·02 i.e. 12·9 c.c. strong acid: 87·1 c.c. water), and the green-blue acid layer is separated and neutralized with amomonia and shaken with more ether, which then contains in solution phytochlorin e, the derivative of chlorophyll a. The phytochlorin e gives to the ether an olive-green colour.

The ether layer remaining in the funnel, after the separation of the green-blue acid layer, is now extracted with 10 c.c. of 12 % hydrochloric acid (sp. gr. 1·06 i.e. 38·1 c.c. strong acid: 61·9 c.c. water). The green acid solution so obtained is diluted with water and shaken with ether which then becomes coloured red and contains phytorhodin g, the derivative of chlorophyll b.

If the phyllins are acted upon by mineral acids, they lose their magnesium in the same way as the chlorophyllins, and the series of substances obtained in this way are termed porphyrins. Thus aetiophyllin will give aetioporphyrin $C_{31}H_{36}N_4$ (see Scheme 1, p. 35).

The derivatives of chlorophyll which are free from magnesium, such as phaeophytin, phytochlorin phytorhodin, the various porphyrins, etc. combine readily with the acetates of some metals such as copper, zinc and iron, and they form intensely coloured, stable compounds. The change of colour is so noticeable that the smallest traces of certain metals can be detected in this way. Hence it is very difficult to prepare the magnesium-free chlorophyll unless the reagents are perfectly pure and all contact with certain metals is avoided.

Expt. 26. *Substitution of copper for magnesium in chlorophyll.* 2 c.c. of an ether solution of chlorophyll are shaken with a little 20 % hydrochloric acid (sp. gr. 1·10 i.e. 62·4 c.c. strong acid: 37·6 c.c. water), and then washed with water in a separating funnel. If the ether tends to evaporate and deposit phaeophytin in the funnel, a little more ether should be added. In this way is produced in ether solution the magnesium-free chlorophyll derivative, phaeophytin. The solution is evaporated down on a water-bath, and the residue dissolved in 5 c.c. of alcohol. The solution is olive-green in colour. This is heated and a grain of copper acetate or zinc acetate is added. The colour changes back to a brilliant green, but without fluorescence (if all the chlorophyll has been converted into phaeophytin).

From the results of these recent investigations, it is now possible to write formulae for the two chlorophylls as follows:

chlorophyll a ($C_{32}H_{30}O\ N_4Mg$) ($COOCH_3$) ($COOC_{20}H_{39}$)
chlorophyll b ($C_{32}H_{28}O_2N_4Mg$) ($COOCH_3$) ($COOC_{20}H_{39}$)

from which it will be seen that the phytol component amounts to one-third of the weight of the chlorophyll. The structural formula for chlorophyll is not completely known, but there is evidence that it contains four pyrrole rings (cp. the pyrrolidine alkaloids, p. 175).

From the analyses of chlorophylls from different plants, it was found that the phytol content varied, and plants which yielded little phytol most readily produced "crystalline chlorophyll," a form of the pigment which has been known for some considerable time to previous workers. The Cow Parsnip (*Heracleum Sphondylium*), Hedge Woundwort (*Stachys sylvatica*) and Hemp-nettle (*Galeopsis Tetrahit*) are plants which readily give crystalline chlorophyll. In this connexion it has been suggested that the chlorophyll in plants is accompanied by an enzyme, chlorophyllase, which, in alcoholic media, brings about alcoholysis of the chlorophyll, and replaces the phytyl by the ethyl radicle. The products, formerly known as crystalline chlorophyll, are now termed chlorophyllides:

$$(C_{32}H_{30}ON_4Mg)(COOCH_3)(COOC_{20}H_{39}) + C_2H_5OH$$
$$= C_{20}H_{39}OH + (C_{32}H_{30}ON_4Mg)(COOCH_3)(COOC_2H_5).$$
$$\text{Phytol} \qquad \text{Ethyl chlorophyllide}$$

Similar chlorophyllides are produced by other alchohols. In aqueous solutions chlorophyllase brings about hydrolysis and the free acid chlorophyllide is formed (see Scheme 2, p. 35):

$$(C_{32}H_{30}ON_4Mg)(COOCH_3)(COOC_{20}H_{39}) + H_2O$$
$$= C_{20}H_{39}OH + (C_{32}H_{30}ON_4Mg)(COOCH_3)(COOH).$$
$$\text{Chlorophyllide}$$

Chlorophyllase is a very stable enzyme; it is not even destroyed by boiling in alcohol for a short time, but if leaves are boiled in water, the enzyme is destroyed.

Expt. 27. *Microscopic examination of ethyl chlorophyllide.* Prepare sections of fresh *Heracleum* leaves and mount them in a drop of 90 % alcohol. Leave the slide under a bell-jar containing a dish of alcohol. The section slowly dries in the course of half a day or a day. It is then examined under the microscope when there will be observed the characteristic triangular and hexagonal crystals of ethyl chlorophyllide (crystalline chlorophyll).

Expt. 28. *Production of methyl chlorophyllide in the leaf.* Sections may be used as in the preceding experiment, or a piece of a leaf may be employed. In the latter case a test-tube with 4 c.c. of 75 % methyl alcohol is taken and 1 gm. of fresh leaf is added to it. The leaf first becomes a darker green and then during the course of a few hours becomes yellowish. On holding the leaf to the light there can be observed with the naked eye a number of black points. If sections of the leaf be cut and examined under the microscope, these spots appear as aggregates composed of rhombohedral crystals, occurring in certain cells.

Expt. 29. *Extraction of ethyl chlorophyllide.* Two grams of dry *Heracleum* leaf-powder are left for a day in a test-tube containing 6 c.c. of 90 % alcohol. The extract is then filtered through a small porcelain funnel and the powder on the filter washed with a little acetone. The filtrate is mixed with an equal quantity of ether, and then with some water. The ether solution is transferred to a separating funnel and thoroughly washed with water, and then concentrated on a water-bath to $\frac{1}{2}$ or 1 c.c., and 3 c.c. of petrol ether are added. On standing, the ethyl chlorophyllide is precipitated in the form of crystalline aggregates. It is freed from yellow pigments by shaking with a *little* ether, and can be further purified by redissolving in ether and precipitating again with petrol ether.

Expt. 30. *The action of chlorophyllase.* Fresh leaves of a species rich in chloro-phyllase, e.g. *Heracleum* or *Galeopsis*, are finely divided and put in a 70 % acetone solution, 3 c.c. of solution being used for every gram of leaf. The chlorophyll, by means of the chlorophyllase, is hydrolyzed into phytol and the acid chlorophyllide. This can be demonstrated after about a quarter of an hour if the solution is diluted with water, transferred to ether and shaken with 0·05 % sodium hydroxide. The sodium hydroxide takes up more colouring matter the further the enzyme action has progressed.

Expt. 31. *The destruction of chlorophyllase.* If fresh leaves of a species rich in chlorophyllase are first steeped in boiling water for a few minutes before they are placed in the acetone solution, unaltered chlorophyll is extracted which does not react with dilute alkali.

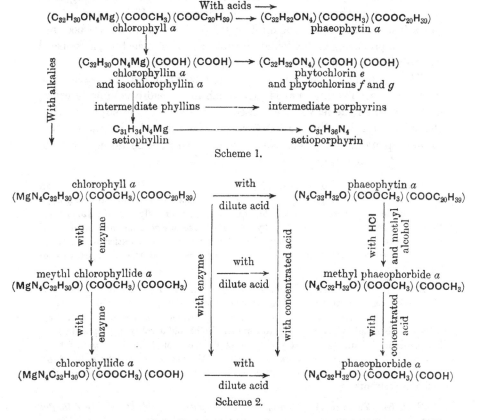

With acids $\longrightarrow$

$(C_{32}H_{30}ON_4Mg)(COOCH_3)(COOC_{20}H_{39}) \longrightarrow (C_{32}H_{32}ON_4)(COOCH_3)(COOC_{20}H_{39})$
chlorophyll a phaeophytin a

$(C_{32}H_{30}ON_4Mg)(COOH)(COOH) \longrightarrow (C_{32}H_{32}ON_4)(COOH)(COOH)$
chlorophyllin a phytochlorin e
and isochlorophyllin a and phytochlorins f and g

intermediate phyllins $\longrightarrow$ intermediate porphyrins

$C_{31}H_{34}N_4Mg \longrightarrow C_{31}H_{36}N_4$
aetiophyllin   aetioporphyrin

With alkalies

Scheme 1.

chlorophyll a with   phaeophytin a
$(MgN_4C_{32}H_{30}O)(COOCH_3)(COOC_{20}H_{39})$ dilute acid $(N_4C_{32}H_{32}O)(COOCH_3)(COOC_{20}H_{39})$

with enzyme   with enzyme   with concentrated acid   with HCl and methyl alcohol

meythl chlorophyllide a with   methyl phaeophorbide a
$(MgN_4C_{32}H_{30}O)(COOCH_3)(COOCH_3)$ dilute acid $(N_4C_{32}H_{32}O)(COOCH_3)(COOCH_3)$

with enzyme   with concentrated acid

chlorophyllide a with   phaeophorbide a
$(MgN_4C_{32}H_{30}O)(COOCH_3)(COOH)$ dilute acid $(N_4C_{32}H_{32}O)(COOCH_3)(COOH)$

Scheme 2.

By treatment with acids, magnesium is removed from the chlorophyl-lides, with the production of the corresponding phaeophorbides. Thus methyl chlorophyllide a gives methyl phaeophorbide a, etc. (see Scheme 2, above).

It has been previously mentioned that water-free solvents, such as acetone, ether and benzene, in which pure extracted chlorophyll is soluble, will not extract the pigment from thoroughly dried leaves, but if a little water is added, it readily goes into solution. From fresh leaves also these solvents can extract the pigments.

As an explanation of the above phenomena, it has been suggested that chlorophyll in the chloroplastid is in the colloidal state, and that, when water is added to the dried leaf, a solution of mineral salts in the leaf is formed which alters the colloidal condition of the chlorophyll and makes it soluble. This view is supported by the fact that if a colloidal solution of chlorophyll in water, made from the pure extracted pigment, is shaken with ether, the ether remains colourless. If, however, a little salt solution is added and the mixture shaken, the ethereal layer becomes green. In preparing the colloidal solution the solvent, acetone, is replaced by the medium, water, in which chlorophyll is insoluble.

The condition of chlorophyll is altered by plunging the leaves into boiling water. The pigment is then much more readily soluble in ether, etc., even when the leaves are subsequently dried. It is supposed that the chlorophyll has diffused out from the plastids, and is in true solution in accompanying waxy substances which have become liquid owing to change of temperature.

Expt. 32. *Preparation of a colloidal solution of chlorophyll.* Take 10 c.c. of an acetone extract of chlorophyll (Expt. 19) and pour this acetone solution into a large volume of distilled water (100 c.c.), the liquid being continually stirred. This operation can be most conveniently done by taking the acetone solution in a pipette and allowing it to run out of the pipette while the latter is used as a stirring rod in the water. Note the change in colour to a purer green, and the disappearance of fluorescence.

Expt. 33. *To demonstrate the difference between a true and a colloidal solution of chlorophyll.* Evaporate 10 c.c. of an acetone extract (Expt. 19) to complete dryness and test its solubility in ether, petrol ether and benzene. It is soluble in all three solvents. Now add these solvents to some of the colloidal solution prepared in the last experiment, and note that the chlorophyll does not dissolve in any of these solvents. If, however, some salt solution, e.g. a little magnesium sulphate, be added, the chlorophyll is precipitated from its colloidal state and is now soluble in ether and other solvents.

Expt. 34. *To show that chlorophyll in the plant is probably in the colloidal condition.* Some nettle powder is carefully dried, e.g. by keeping it at 30–40° C. in an oven, and then further drying in a vacuum desiccator over sulphuric acid. Small quantities of this dry powder are put in test-tubes, and different pure water-free substances such as acetone, ether, benzene and absolute alcohol are added. Note that these solvents are not coloured by the chlorophyll. It can be demonstrated that the extracted pigment is easily soluble in any of these substances. Repeat the experiment with nettle powder moistened with a few drops of water, and note that the solvents are immediately coloured.

Expt. 35. *Pure solvents are able to extract chlorophyll from fresh leaves.* Crush 10 gms. of fresh leaves of nettle, horse-chestnut or elder in a mortar with some clean sand, and put the crushed material on a filter-paper in a porcelain funnel. Add 20 c.c. of pure acetone and suck it through by means of a water-pump. Repeat this several times. The pure solvent is here able to extract the pigment.

Expt. 36. *Treatment of fresh leaves with boiling water changes the condition of the chlorophyll.* Dry a quantity of leaves which have been put in boiling water and examine their solubility as in Expt. 34. Note that the chlorophyll in this powder is soluble in pure solvents.

There is finally another change which chlorophyll can undergo, namely that of allomerization, which takes place in alcoholic solution. The characteristic of allomerized chlorophyll is that it does not give the brown phase when treated with alkali (see Expt. 23). Allomerization is accelerated in alkaline solution but inhibited by small quantities of acid.

Expt. 37. *To demonstrate that allomerized chlorophyll does not give the brown phase test.* Dissolve a little crude chlorophyll, obtained by evaporating an ether solution, in absolute alcohol. To a sample of this add a little alkali, and perform the phase test, from time to time, till at last the brown phase no longer appears.

CONNEXION OF CHLOROPHYLL WITH FORMALDEHYDE.

In addition to the above, another chemical property of chlorophyll of great interest, is that connected with the production of formaldehyde. Those investigators, who have sought to confirm the formaldehyde hypothesis of carbon assimilation, have based their evidence on tests for formaldehyde both in the plant and in chlorophyll-containing systems outside the plant. By exposing films, or solutions, of chlorophyll to light in presence of carbon dioxide, they have detected formaldehyde as a result (Usher and Priestley, 5).

The most recent investigations (Jörgensen and Kidd, 2) have shown that the experimental evidence is at present inadequate to support the hypothesis, since formaldehyde arises from chlorophyll itself in the absence of carbon dioxide.

In this later work (Jörgensen and Kidd, 2) on the behaviour of extracted chlorophyll in light, use has been made of a colloidal solution (see p. 36) of pure chlorophyll (chlorophylls a and b) for experimental work. The solution for this purpose must be prepared from pure chlorophyll which has been tested and shown to be free from yellow pigments, since the latter absorb oxygen and may confuse the issue of the experiment. The pure chlorophyll is prepared by extracting dried nettle leaves with 80–85 % acetone in the usual way and transferring to petrol ether (p. 29). The petrol ether extract is then washed with 80 % acetone to remove colourless impurities, and with 80 % methyl alcohol to remove xanthophyll. Finally all traces of acetone and methyl alcohol are removed by washing with water. This renders the chlorophyll insoluble in petrol ether, since it is only soluble in this solvent if traces of other solvents are present. Hence the pigment is precipitated out as a fine suspension, leaving the carotin in solution. The chlorophyll is filtered off through powdered talc, taken up in ether, reprecipitated by petrol ether and finally obtained as a blue-black micro-crystalline substance. The colloidal solution or sol is made by dissolving 0·4 gm. of pure chlorophyll in 3 c.c. of absolute alcohol and pouring into 300 c.c. of distilled water.

The advantage of using such a solution is that the experimental conditions, in all probability, approach more nearly to the conditions in the plant, and reactions with other substances take place more readily than when the chlorophyll is used as a film. The use of pure, instead of crude, chlorophyll is also important as by this means it is possible to determine the changes taking place in chlorophyll itself without complications arising from the accompanying impurities. The discordant results of various workers on this subject are doubtless due to the employment of crude chlorophyll. Ethyl alcohol is the best solvent for preparing the sol since it does not produce formaldehyde when exposed to light under ordinary circumstances in glass vessels. Methyl alcohol and acetone should be avoided as they themselves either contain or give rise to formaldehyde.

The chlorophyll sol is electro-negative. It is stablized by weak alkalies, but precipitated by weak acids.

Working with such a colloidal solution the results may be summarized as follows.

When a chlorophyll sol is exposed to light in an atmosphere of nitrogen in a sealed tube, no apparent change takes place in the chlorophyll, and no formaldehyde is produced.

When exposed in an atmosphere of carbon dioxide in a sealed tube, the chlorophyll rapidly turns yellow- or brown-green. In the case of sols of high concentration, the colour-change is preceded by precipitation of the pigment. The same change takes place in the dark, only more slowly. No formaldehyde is produced, and no absorption of carbon dioxide could be detected. The yellow product has been shown to be the magnesium-free derivative, phaeophytin, which is produced from the pigment by the action of acids. The changes observed are explained by the fact that the carbon dioxide, acting as a weak acid, first precipitates the sol, if concentrated, and then acts, like other weak acids, on the chlorophyll, producing phaeophytin. If the solution is kept neutral by addition of sodium bicarbonate, there is no colour change. The identity of phaeophytin was shown by the spectrum and by the restoration of colour on adding a trace of copper acetate.

When exposed to light, and the atmosphere in the sealed tube is replaced by oxygen or air, the chlorophyll turns yellow- or brown-green as before and then bleaches. The change of colour from green to yellow or brown is again due to the formation of phaeophytin, this being brought about by the presence of an acid substance, which is produced during bleaching, and increases throughout the process. Formaldehyde can be detected in a very slight amount during bleaching, but is formed in much greater quantity after bleaching is complete.

It is suggested that the formaldehyde is produced by the oxidation and breaking down of the phytol component of the chlorophyll:

$$CH_3-CH-CH-CH-CH-CH-CH-CH-C=C-CH_2OH$$
$$\quad\ \ CH_3\ \ CH_3\ \ CH_3\ \ CH_3\ \ CH_3\ \ CH_3\ \ CH_3\ \ CH_3\,CH_3$$

There is no reason for ascribing to any of the above reactions any part in carbon assimilation. There is at present no hypothesis, supported by satisfactory evidence, as to the process of carbon assimilation.

Expt. 38. *Detection of formaldehyde as a product of oxidation of chlorophyll.* Extract 2 gms. of dried nettle leaf powder with 20 c.c. of 80 % acetone and transfer it to petrol ether as in Expt. 19. Then shake the petrol ether extract four or five times with an equal volume of 80 % acetone to remove colourless impurities. Next the petrol ether extract is similarly shaken up with 80 % methyl alcohol which removes the xanthophyll. This should be repeated until the methyl alcohol is colourless. The petrol ether is finally washed repeatedly with water to remove traces of acetone and methyl alcohol. The chlorophyll is in time precipitated as a fine suspension, being insoluble in pure petrol ether. This suspension is filtered through either kieselguhr or powdered talc on a small porcelain filter. The chlorophyll is extracted from the powder on the filter with as small a quantity as possible of absolute alcohol. This alcoholic solution is then poured, with constant stirring, into 100 c.c. of distilled water by which means a colloidal solution of chlorophyll is obtained.

The test to be employed for formaldehyde is as follows (Schryver, 4). To 10 c.c. of the liquid to be tested add 2 c.c. of a 1 % solution (freshly made) of phenylhydrazine hydrochloride, 1 c.c. of a 5 % solution (freshly made) of potassium ferricyanide and 5 c.c. of concentrated hydrochloric acid. If formaldehyde is present a pink to magenta colour is developed, either deep or pale, according to the quantity of formaldehyde.

The reaction is due to the formation of a condensation product of formaldehyde and phenylhydrazine, and this compound, on oxidation, yields a weak base forming a coloured salt with concentrated hydrochloric acid. The salt is readily dissociated again on dilution of the solution.

Two modifications (Schryver, 4) can be adopted in applying this test. First, in testing for formaldehyde in pigmented solutions, the following course can be pursued. The reaction mixture, after addition of phenylhydrazine, ferricyanide and hydro-chloric acid, is diluted with water, and ether is added in a separating funnel. The hydrochloride of the chromatogenic base is dissociated and the base is taken up by the ether. The aqueous solution is run off, and on addition of strong hydrochloric acid to the ether, the base passes into the acid as a coloured hydrochloride again. By using a small quantity of acid, the sensitiveness of the test is increased, since the colour is now distributed through a small quantity of liquid only.

The second modification consists in warming the solution to be tested for a short time with the phenylhydrazine hydrochloride before adding the other reagents. In this way, formaldehyde can also be detected if it should be in a polymerized form.

As a control, 10 c.c. of the colloidal solution of chlorophyll should be tested, using both the above modifications. The remainder of the solution should be exposed to sunlight (or the light from either an arc or mercury vapour lamp) in a loosely corked vessel, until it is completely bleached. The bleached solution, on testing, will be found to give a positive test for formaldehyde.

The Yellow Plastid Pigments.

These have already been mentioned in connexion with the leaf pigments (pp. 29 and 30). In addition, however, they have a further significance in that they constitute the pigments, located in plastids, of most yellow and orange flowers and fruits. Sometimes also they occur in other organs, i.e. root of Carrot (carotin).

Carotin, $C_{40}H_{56}$, is an unsaturated hydrocarbon. It crystallizes in lustrous rhombohedra which are orange-red by transmitted and blue by reflected light. It is readily soluble in chloroform, benzene and carbon bisulphide, but with difficulty in petrol ether and ether.

One of its most characteristic properties is that it readily undergoes oxidation in air, and becomes bleached. With concentrated sulphuric acid it gives a deep blue colour.

Xanthophyll, $C_{40}H_{56}O_2$, also forms yellow crystals with a blue lustre. It is soluble in chloroform and ether, but insoluble in petrol ether. It is more soluble than carotin in methyl alcohol. It gives a blue colour with sulphuric acid, and also oxidizes in air with bleaching.

The separation of the two pigments (see Expt. 22) is based on the fact that in a mixture of petrol ether and methyl alcohol containing a little water, the carotin passes entirely into the petrol ether, whereas the greater part of the xanthophyll remains in the methyl alcohol layer.

REFERENCES

BOOKS

1. **Willstätter, R.**, und **Stoll, A.** Untersuchungen über Chlorophyll. Methoden und Ergebnisse. Berlin, 1913.

PAPERS

2. **Jörgensen, I.**, and **Kidd, F.** Some Photochemical Experiments with Pure Chlorophyll and their Bearing on Theories of Carbon Assimilation *Proc. R. Soc.*, 1917, B Vol. 89, pp. 342–361.

3. **Jörgensen, I.**, and **Stiles, W.** Carbon Assimilation. A Review of Recent Work on the Pigments of the Green Leaf and the Processes connected with them. *New Phytologist*, Reprint, No. 10. London, 1917.

4. **Schryver, S. B.** The Photochemical Formation of Formaldehyde in Green Plants. *Proc. R. Soc.*, 1910, B Vol. 82, pp. 226–232.

5. **Spoehr, H. A.** Photosynthesis. New York, 1926.

6. **Stiles, W.** Photosynthesis. London, 1925.

7. **Usher, F. L.**, and **Priestley, J. H.** A Study of the Mechanism of Carbon Assimilation in Green Plants. I. *Proc. R. Soc.*, 1906, B Vol. 77, pp. 369–376. II. *Ibid.* 1906, B Vol. 78, pp. 318–327. III. *Ibid.* 1912, B Vol. 84, pp. 101–112.

CHAPTER V

CARBOHYDRATES

THE carbohydrates which occur in plants may be classified as follows:

Monosaccharides......
- Pentoses, $C_5H_{10}O_5$—Arabinose, xylose.
- Methyl pentoses, $C_5H_9O_5 \cdot CH_3$—Rhamnose, isorhamnose.
- Hexoses, $C_6H_{12}O_6$—Glucose, galactose, mannose, laevulose.

Disaccharides ....
- Sucrose, maltose, $C_{12}H_{22}O_{11}$.

Trisaccharides.........
- Raffinose and others.

Tetrasaccharides......
- Stachyose.

Polysaccharides ......
- Pentosans, $(C_5H_8O_4)_n$—Araban, xylan.
- Starches, $(C_6H_{10}O_5)_n$—Starch, dextrin, inulin.
- Mannans, galactans, gums, mucilages, pectic substances.
- Celluloses, $(C_6H_{10}O_5)_n$.

The carbohydrates are widely distributed in plants and form most important parts of their structure. Those most commonly found are: cellulose, starch, pentosans, dextrin, glucose, sucrose, laevulose, and maltose. Other sugars, especially trisaccharides, are known in addition to those mentioned above, but they are somewhat restricted and specific in their distribution.

As in the case of the proteins, so with the carbohydrates, the molecules of the more simple and soluble crystalline compounds, such as the monosaccharides, are synthesized into more complex molecules which exist, either in the colloidal (dextrin), or insoluble state (starch, cellulose). The last-mentioned build up parts of the solid structure of the plant. The resolution of the solid complex substances into simple ones is known in many instances to be brought about in the plant by enzymes, and it is highly probable that the synthesis of the complex from the simple is also controlled by these enzymes.

The most commonly occurring sugars in plants are glucose, laevulose sucrose and maltose : sucrose is hydrolyzed by the enzyme, invertase, into one molecule of glucose and one molecule of laevulose : maltose by the enzyme, maltase, into two molecules of glucose. Both invertase and maltase are widely distributed. The connexion between various sugars and photosynthesis, and their inter-relationships with each other in the leaves, are reserved for another section.

Of the polysaccharides, cellulose is universally distributed in higher plants and constitutes the greater part of the cell-walls. The pentosans, galactans and mannans also, but to a lesser degree, are components of their structure. Starch, in addition, is very widely distributed : it is converted by the enzyme, diastase, into dextrin and maltose, and possibly the same enzyme also controls its synthesis. In some plants no starch is formed, and its place in metabolism is taken by inulin or cane-sugar.

The various carbohydrates will first be dealt with in detail, and later their inter-relationships will be considered.

MONOSACCHARIDES.

These are termed tetroses, pentoses or hexoses according to the number of carbon atoms in the molecule. They contain primary $(- CH_2OH)$ or secondary $(= CHOH)$ alcohol groups, and either an aldehyde $(- CHO)$ group, as in glucose, or a ketone $(= C = O)$ group, as in laevulose. They are, as a class, white crystalline substances, soluble in water and aqueous alcohol, but insoluble in ether, acetone and many other organic solvents. They are capable of certain characteristic chemical reactions which form a basis for their detection and estimation. One of the most important is that connected with the aldehyde and ketone groups, owing to which they act as reducing agents, being themselves oxidized. The reducing action usually employed is that which takes place with copper salts in hot alkaline solution, whereby cuprous oxide is formed. Hence they are termed " reducing " sugars. Another important reaction is the formation of crystalline osazones (only in the case of sugars with aldehyde or ketone groups), which, by virtue of their melting points and charac-teristic crystalline forms, constitute, in several cases, valuable tests for the presence of sugars.

A reaction exhibited by many of the monosaccharides is that of forming a coloured product when heated with a phenol in presence of a strong acid. The reaction is due to the formation of a furfural compound (see p. 46), and the colour depends on the particular sugar and phenol used. Thus, with strong hydrochloric acid and orcinol, the colour is violet-blue for pentoses and orange-red for hexoses; with the same acid and phloroglucinol, the colour is red in both cases; with α-naphthol and strong sulphuric acid, the colour is purple in all cases. A variation of this reaction provides a distinction between a ketone and an aldehyde sugar. Thus, if hydrochloric acid diluted with its own volume of water is used, a red colour is produced with resorcinol and a ketone sugar, e.g. laevulose (Seliwanoff's reaction). With an aldehyde sugar, e.g. glucose, the colour is produced only by using concentrated acid.

PENTOSES, METHYL PENTOSES.

The pentoses contain five carbon atoms, and have the general formula $C_5H_{10}O_5$. They are said to be present in the free state to some extent in leaves (Davis and Sawyer, 12). In plants they occur chiefly, however, as condensation products formed with elimination of water. These products are termed the pentosans, and are widely distributed; on hydrolysis they yield pentoses again. The various gums found in plants consist largely of pentosans, and the pectins also contain pentose groups; both consequently yield pentoses on hydrolysis (see pp. 63 and 66). A pentose is also a component of plant nucleic acid (see p. 141). It has recently been shown (Spoehr, 33) that the metabolism of some succulent plants (Cactaceae) is especially favourable to the production of pentoses. By condensation, pentosan-mucilage is formed and this has the water-retaining properties characteristic of succulents. (See Appendix, p. 183.)

If we examine the structural formula of a pentose, as for example arabinose:

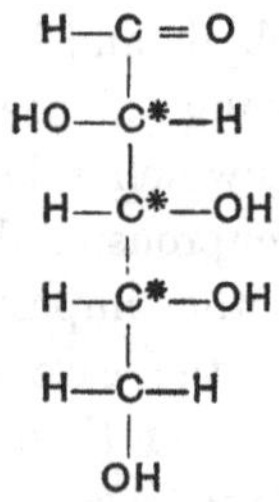

$$
\begin{array}{c}
\text{H—C = O} \\
| \\
\text{HO—C*—H} \\
| \\
\text{H—C*—OH} \\
| \\
\text{H—C*—OH} \\
| \\
\text{H—C—H} \\
| \\
\text{OH}
\end{array}
$$

we see that each of the three carbon atoms marked * is united to four different atoms or groups of atoms. Each of these carbon atoms is therefore asymmetric, and, with regard to it, there are two possible isomers (see p. 10, Cole, 5, for stereoisomerism). It will be found on examination that there are eight possible isomers of the formulae given above:

```
      CHO               CHO               CHO               CHO
       |                 |                 |                 |
  HO—C—H            H—C—OH            HO—C—H            H—C—OH
       |                 |                 |                 |
  HO—C—H            H—C—OH            H—C—OH            HO—C—H
       |                 |                 |                 |
  HO—C—H            H—C—OH            HO—C—H            H—C—OH
       |                 |                 |                 |
     CH2OH             CH2OH             CH2OH             CH2OH
    l-Ribose          d-Ribose          l-Xylose          d-Xylose


      CHO               CHO               CHO               CHO
       |                 |                 |                 |
  H—C—OH            HO—C—H            H—C—OH            HO—C—H
       |                 |                 |                 |
  HO—C—H            H—C—OH            H—C—OH            HO—C—H
       |                 |                 |                 |
  HO—C—H            H—C—OH            HO—C—H            H—C—OH
       |                 |                 |                 |
     CH2OH             CH2OH             CH2OH             CH2OH
   l-Arabinose       d-Arabinose       l-Lyxose          d-Lyxose
                                       unknown
```

Of these only seven have been isolated. The pentoses which occur in plants are l-arabinose, d-xylose[1] and d-ribose. The two former, however, are known almost solely as condensation products, pentosans, in gums, woody tissue, etc.; the latter only as a component of nucleic acid. The pentoses form osazones (see p. 51 for reactions and composition).

Arabinose. This sugar occurs as the pentosan, araban, in various gums, such as Cherry Gum, Gum Arabic, etc. (see p. 46).

Some of the properties and reactions of the pentoses are demonstrated in the following experiments.

Expt. 39. *Tests for arabinose.* For reactions *a–e* use a 1 % solution of arabinose: for reaction *f* a 0·2 % solution.

If pure arabinose is not available, a solution for tests *a*, *b* and *c* can be prepared from gum arabic. Boil 5 gms. of the gum in 100 c.c. of water with 10 c.c. of concentrated hydrochloric acid for 5 minutes and then neutralize to litmus with alkali. Such a solution is only suitable for the specific tests for arabinose, since it also contains galactose (see p. 63). For tests *a*, *b* and *c* small pieces of solid gum arabic may even be used.

(*a*) Heat a few c.c. of the sugar solution in a test-tube with about half its volume of concentrated hydrochloric acid. In the mouth of the test-tube place a piece of filter-paper soaked with aniline acetate (made by mixing equal quantities of aniline, water and glacial acetic acid). A pink colour will be produced in the paper. This is

[1] Known formerly as l-xylose.

due to the fact that furfural is formed by the action of the acid on the pentose, and the furfural then gives a red colour with aniline acetate solution :

Arabinose   Furfural

This reaction, however, is also given by the hexoses but to a much less extent.

(*b*) Warm a few c.c. of the sugar solution with an equal volume of concentrated hydrochloric acid in a test-tube, and add a small quantity of phloroglucinol. A bright red coloration is produced.

(*c*) To a few c.c. of the sugar solution in a test-tube add an equal quantity of concentrated hydrochloric acid, and then a little solid orcinol. Divide the solution into two equal portions. Heat one portion. The solution becomes bluish changing to reddish-violet and finally deposits a blue precipitate. To the other portion, after heating for a time, add a few drops of 10 % ferric chloride solution. A deep green colour is at once produced. On the addition of a little amyl alcohol, the green colour will be extracted by the alcohol.

(*d*) *a*-Naphthol reaction. Add to a little of the sugar solution a few drops of a 1 % solution of *a*-naphthol in alcohol. Mix the two solutions and then run in about 5 c.c. of concentrated sulphuric acid down the side of the test-tube. A violet coloration is produced at the junction of the two liquids. The coloration is due to a condensation product of *a*-naphthol with furfural, the latter being formed by the action of the acid on the carbohydrate. This reaction is likewise given by laevulose and cane-sugar (since it yields laevulose, see p. 54), and less strongly by glucose and maltose; also by some proteins which contain a carbohydrate group.

(*e*) Boil a little of the arabinose solution with a few drops of Fehling's solution. Reduction will take place.

(*f*) Make the osazone of arabinose following the instructions given for glucosazone (see p. 50).

A solution of arabinose which will give the pentose reactions can also be obtained by hydrolysis of Cherry Gum. The gum oozes from the bark of various species of *Prunus*, such as the Cherry (*Prunus Cerasus*) and the Bird Cherry (*P. Padus*).

Expt. 40. *Preparation of arabinose solution from Cherry Gum.* The gum is heated, on a water-bath in a round-bottomed flask fitted with an air condenser[1], with dilute sulphuric acid (1 pt. by wt. of gum: 7 pts. by wt. of 4 % sulphuric acid) for about 5 hours. The solution is then neutralized with calcium carbonate and filtered. Perform the tests *a*, *b* and *c* of Expt. 39 on the solution. A positive result is obtained in each case. Since the solution contains other sugars as impurities, it cannot con-

---

[1] i.e. a wide piece of glass tubing about 3 ft. long passing through the cork.

clusively be used for tests *d*, *e* and *f*. If a considerable quantity of gum is available, crystallization of arabinose should be attempted by concentrating the aqueous sugar solution, extracting this with 90 % alcohol and again concentrating in a desiccator (see p. 55). If a very small quantity of gum only is available, the tests *a*, *b* and *c* should be performed directly on a small piece of the gum in a test-tube.

A purer preparation of arabinose, which may be used for all the tests of Expt. 39, can be obtained by the hydrolysis of araban (see Expt. 48).

Xylose. This sugar occurs very widely distributed in woody tissue as the pentosan, xylan (see p. 56). A solution of xylose which will give the pentose reactions can be obtained from the hydrolysis of straw, or the presence of xylan giving the pentose reactions can be directly demonstrated in straw, bran or sawdust (see Expt. 49).

A purer solution of xylose can be obtained from the hydrolysis of xylan (see Expt. 51).

When xylose is oxidized with bromine, it yields xylonic acid which has a characteristic cadmium salt. The formation of this salt is used as a method for identifying the sugar (see Expt. 51).

The methyl pentoses are pentoses in which one of the hydrogen atoms of the CH_2OH group is replaced by the methyl group, CH_3.

Rhamnose, $C_5H_9O_5CH_3$, occurs as the constituent of many glucosides (see pp. 113, 159).

HEXOSES.

If we examine the structural formula for a hexose, such as glucose:

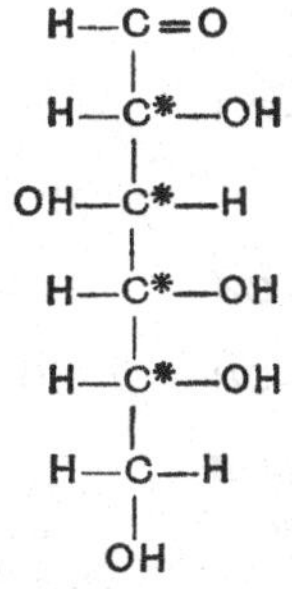

$$
\begin{array}{c}
H-C=O \\
| \\
H-C^*-OH \\
| \\
OH-C^*-H \\
| \\
H-C^*-OH \\
| \\
H-C^*-OH \\
| \\
H-C-H \\
| \\
OH
\end{array}
$$

we see that there are four carbon atoms marked * which are united to four different groups of atoms. It will be found in this case that there are sixteen possible isomers, as against eight for pentose:

```
      CHO              CHO              CHO              CHO
       |                |                |                |
   H—C—OH          HO—C—H           HO—C—H           H—C—OH
       |                |                |                |
   H—C—OH          HO—C—H           H—C—OH           HO—C—H
       |                |                |                |
  HO—C—H           H—C—OH           HO—C—H           H—C—OH
       |                |                |                |
  HO—C—H           H—C—OH           HO—C—H           H—C—OH
       |                |                |                |
    CH₂OH            CH₂OH            CH₂OH            CH₂OH
  l-Mannose        d-Mannose        l-Glucose        d-Glucose
```

```
      CHO              CHO              CHO              CHO
       |                |                |                |
   H—C—OH          HO—C—H           HO—C—H           H—C—OH
       |                |                |                |
  HO—C—H           H—C—OH           HO—C—H           H—C—OH
       |                |                |                |
   H—C—OH          HO—C—H           H—C—OH           HO—C—H
       |                |                |                |
  HO—C—H           H—C—OH           HO—C—H           H—C—OH
       |                |                |                |
    CH₂OH            CH₂OH            CH₂OH            CH₂OH
  l-Idose          d-Idose¹         l-Gulose         d-Gulose²
```

```
      CHO              CHO              CHO              CHO
       |                |                |                |
  HO—C—H           H—C—OH           H—C—OH           HO—C—H
       |                |                |                |
   H—C—OH          HO—C—H           H—C—OH           HO—C—H
       |                |                |                |
   H—C—OH          HO—C—H           H—C—OH           HO—C—H
       |                |                |                |
  HO—C—H           H—C—OH           HO—C—H           H—C—OH
       |                |                |                |
    CH₂OH            CH₂OH            CH₂OH            CH₂OH
  l-Galactose      d-Galactose      l-Talose         d-Talose
```

```
      CHO              CHO              CHO              CHO
       |                |                |                |
  HO—C—H           H—C—OH           H—C—OH           HO—C—H
       |                |                |                |
  HO—C—H           H—C—OH           HO—C—H           H—C—OH
       |                |                |                |
  HO—C—H           H—C—OH           HO—C—H           H—C—OH
       |                |                |                |
  HO—C—H           H—C—OH           HO—C—H           H—C—OH
       |                |                |                |
    CH₂OH            CH₂OH            CH₂OH            CH₂OH
  l-Allose         d-Allose         l-Altrose        d-Altrose
  unknown                           unknown
```

¹ Known formerly as l-Idose.   ² Known formerly as l-Gulose.

Though many of the above sugars have been synthesized artificially, only three are known to occur naturally, i.e. d-glucose (dextrose or grape-sugar), d-mannose and d-galactose.

Since compounds containing asymmetric carbon atoms are optically active, i.e. can rotate a plane of polarized light, it follows that the sugars under discussion are optically active.

Glucose. This substance, which is also known as grape-sugar, is very common and very widely distributed in plants. It occurs in the tissues of leaves, stems, roots, flowers and fruits. It is produced as a result of the hydrolysis of cane-sugar and maltose, and, in all probability, is the first sugar synthesized from carbon dioxide and water. Its synthesis and its relationships to other sugars will be discussed later (see p. 71). It is a white crystalline substance, readily soluble in water and aqueous alcohol, but only slightly soluble in absolute alcohol.

d-glucose is dextro-rotatory.

When either d- or l-glucose is first dissolved in water, it is chemically less active than would be expected of the aldehyde form depicted above. This is explained by assuming that glucose, when first dissolved in water exists in the condition of a lactone (see Appendix, p. 184):

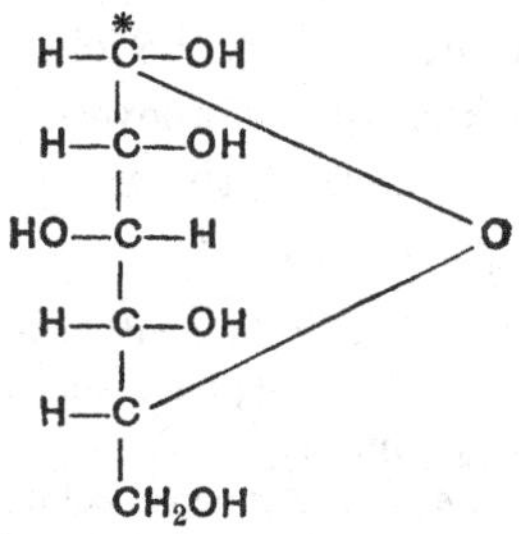

$$
\begin{array}{l}
\overset{*}{H-C}-OH \\
\quad | \\
H-C-OH \\
\quad | \\
HO-C-H \qquad\qquad O \\
\quad | \\
H-C-OH \\
\quad | \\
H-C \\
\quad | \\
CH_2OH
\end{array}
$$

In the above state the carbon atom marked * is also asymmetric so that two forms of glucose are possible, a- and β-glucose:

$$
\begin{array}{l}
H-C-OH \\
\quad | \\
H-C-OH \\
\quad | \\
HO-C-H \qquad O \\
\quad | \\
H-C-OH \\
\quad | \\
H-C \\
\quad | \\
CH_2OH
\end{array}
\qquad\qquad
\begin{array}{l}
HO-C-H \\
\quad | \\
H-C-OH \\
\quad | \\
HO-C-H \qquad O \\
\quad | \\
H-C-OH \\
\quad | \\
H-C \\
\quad | \\
CH_2OH
\end{array}
$$

a-Glucose β-Glucose

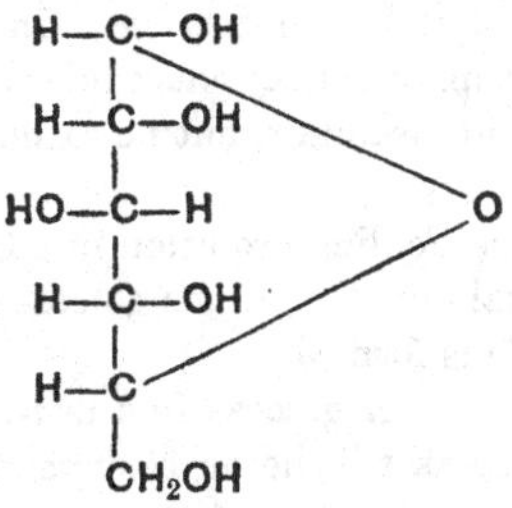

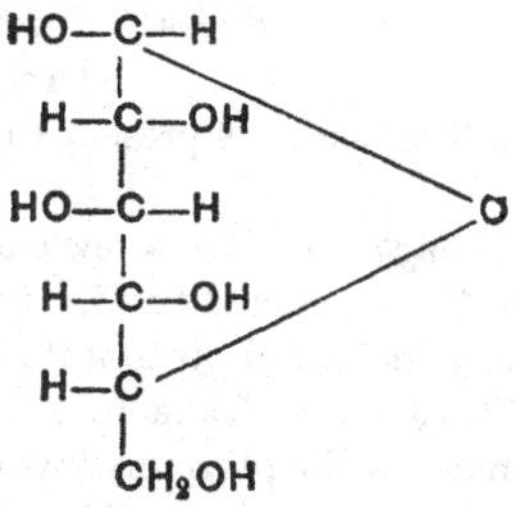

In solution, both the above forms pass by tautomerism into the aldehyde form.

In the plant there are, as will be described later (p. 157), many aromatic and other compounds containing one or more hydroxyl groups. These hydroxyl groups of the aromatic substances are frequently replaced by a glucose (or other sugar) molecule, and such compounds are termed glucosides, as, for instance, salicin, the glucoside of salicylic alcohol which occurs in Willow bark (see p. 167):

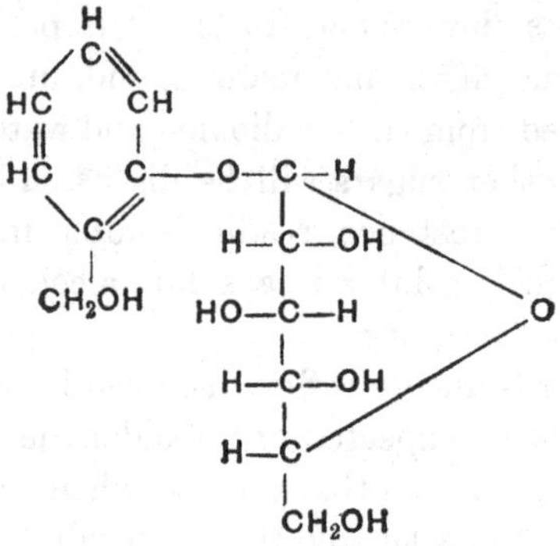

Salicin

These substances, moreover, may be classified either as a- or β-glucosides according to which of the above a or β forms of glucose has combined with the residual part of the compound. Various glucosides will be dealt with in Chaps. VIII and X.

Expt. 41. *Tests for glucose.* Before dealing with the sugars actually isolated from the plant, it is advisable that the following tests and reactions should be performed with pure glucose using a 0·2% solution.

(a) *Moore's test.* Boil a little of the glucose solution with an equal volume of caustic soda solution. A yellow colour is developed which is due to the formation of a condensation product (caramel) of the sugar.

(b) *Trommer's test.* Add a few drops of a 1% copper sulphate solution to 2–3 c.c. of 5% caustic soda solution. A blue precipitate of cupric hydroxide is formed. Add now 2–3 c.c. of the glucose solution, and the precipitate will dissolve. On boiling, the blue colour disappears, and a yellow or red precipitate of cuprous oxide is formed. If only a little sugar is present the blue colour will disappear, but no oxide may be formed.

(c) *Fehling's test.* Boil a few c.c. of freshly made Fehling's solution in a test-tube and note that it is unaltered. Then add an equal quantity of the glucose solution and boil again. A red precipitate of cuprous oxide is formed.

(d) *Osazone test.* Take 10 c.c. of a 0·5% solution of glucose in a test-tube and add as much solid phenylhydrazine hydrochloride as will lie on a sixpenny piece, at least twice as much solid sodium acetate and also 1 c.c. of strong acetic acid

Warm gently until the mixture is dissolved and filter into another test-tube. Then place the tube in a beaker of boiling water for at least $\frac{1}{2}$ hour, keeping the water boiling all the time. Let the test-tube cool slowly, and a yellow crystalline deposit of phenylglucosazone will separate out. Examine this under the microscope and it will be found to consist of fine yellow needles variously aggregated into sheaves and rosettes. Glucosazone melts at 204–205°C.

The osazone reaction takes place as follows:

$$CH_2OH\,(CHOH)_4\,CHO + H_2N\cdot NHC_6H_5 = CH_2OH\,(CHOH)_4\,CH:N\cdot NHC_6H_5 + H_2O.$$
Glucose phenylhydrazone

The phenylhydrazone is very soluble, but if an excess of phenylhydrazine is used, a second hydrazine complex is introduced and an insoluble osazone is formed:

$$CH_2OH\,(CHOH)_3\text{---}C\text{---}CH:N\cdot NHC_6H_5$$
$$\overset{\|}{N}\cdot NHC_6H_5$$

Glucose reacts in this way by virtue of its aldehyde group. Phenylhydrazine hydrochloride does not give an osazone when boiled with glucose unless excess of sodium acetate be added. This acts on the hydrochloride to form phenylhydrazine acetate and sodium chloride.

Galactose. Galactose rarely, if ever, occurs free in plants, though it is fairly widely distributed in the form of condensation products, the galactans, in combination with other hexoses and with pentoses (see p. 62). These galactans form constituents of various gums, mucilages, etc. Agar-agar, which is a mucilage obtained from certain genera of the Red Seaweeds (Rhodophyceae), yields a high percentage of galactose on hydrolysis with acids. Galactose also occurs as a constituent of some glucosides from which it may be derived on hydrolysis. (See Appendix, p. 184.)

One of the most important reactions of galactose is the formation of mucic acid on oxidation with nitric acid. Mucic acid is practically insoluble in water and separates out as a crystalline precipitate on pouring the products of oxidation into excess of water.

Expt. 42. *Preparation of galactose from agar-agar.* Weigh out 50 gms. of agar-agar. Put it into a round-bottomed flask fitted with an air condenser (see p. 46). Add 500 c.c. of 2 % sulphuric acid and heat on a water-bath for 4 hrs. Neutralize the solution with calcium carbonate and filter. Concentrate on a water-bath to a syrup. On standing, crystals of galactose will separate out. Then add a little 50–75 % alcohol and warm gently on a water-bath. By this means much of the dark-coloured product will go into solution and can be poured off leaving the crystalline residue. Take up this residue in a little hot water, boil well with animal charcoal to decolorize the solution and filter. Concentrate again on a water-bath. On cooling, colourless prisms of galactose will separate out.

Expt. **43.** *Oxidation of galactose to mucic acid.* Heat the galactose obtained in
the last experiment with nitric acid (1 gm. galactose to 12 c.c. of nitric acid of sp. gr.
1·15, i.e. 5 pts. of concentrated acid and 12 pts. of water) on a water-bath, until the
liquid is reduced to one-third of its bulk. Then pour the product into excess of
distilled water. On standing (for a day or two), a white sandy microcrystalline precipitate of mucic acid will separate out.

Mannose. Mannose has not been detected free in many plants, but
is widely distributed as condensation products, the mannans, in certain
mucilages and in the cell-walls of the endosperm of various seeds (see
p. 61). From the mannans the sugar can be obtained by hydrolysis.
On adding phenylhydrazine hydrochloride and sodium acetate to a solution of mannose, the phenylhydrazone, which is nearly insoluble in water,
is formed almost immediately and hence constitutes a ready method for
the detection of the sugar.

Laevulose. This sugar, which is also termed fructose, is widely distributed in plants, in the tissues of leaves, stems, fruits, etc. It is formed,
together with glucose, in the hydrolysis by acids of cane-sugar. The
original cane-sugar is dextro-rotatory, whereas laevulose is more laevo-
rotatory than glucose is dextro-rotatory; hence the mixture from the
hydrolysis is laevo-rotatory and is known as invert sugar, the change
being termed inversion. The same hydrolysis is brought about by the
widely distributed enzyme, invertase. The polysaccharide, inulin, also
yields laevulose on acid hydrolysis. Laevulose is a white crystalline substance, soluble in water and alcohol. Unlike glucose, it contains a ketone
instead of an aldehyde group:

$$CH_2OH$$
$$|$$
$$C=O$$
$$|$$
$$HO-C-H$$
$$|$$
$$H-C-OH$$
$$|$$
$$H-C-OH$$
$$|$$
$$CH_2OH$$
d-Fructose

Laevulose reduces Fehling's and other copper solutions. It yields
the same osazone as glucose with phenylhydrazine hydrochloride and
sodium acetate. It also forms an osazone with methylphenylhydrazine
(m.p. 158° C.), a reaction which constitutes a distinction from glucose
since the latter gives no osazone with this substance.

Expt. 44. *Tests for laevulose.* The following tests should be performed with a 0·2 % solution of laevulose in the same way as for glucose (see p. 50).

(*a*) *Moore's test.* A positive result is obtained.

(*b*) *Trommer's test.* A positive result is obtained.

(*c*) *Fehling's test.* Reduction takes place.

(*d*) *Osazone test.* Note that the crystals are identical with those formed from glucose.

(*e*) *a-Naphthol test* (see p. 46). A strong reaction is given.

(*f*) *Seliwanoff's test.* To 5 c.c. of Seliwanoff's solution (prepared by dissolving 0·05 gm. of resorcinol in 100 c.c. of 1 in 2 hydrochloric acid) add a few drops of laevulose solution and boil. A red coloration and a red precipitate are formed. Add a little alcohol and the precipitate forms a red solution (see p. 44).

DISACCHARIDES.

These sugars are formed from the monosaccharides by condensation with elimination of water. By boiling with dilute acids, or by the action of certain enzymes, they are hydrolyzed into monosaccharides. The two most important disaccharides found in plants are maltose and cane-sugar.

Maltose. Maltose or malt-sugar, though it probably occurs in smaller quantities than glucose and laevulose, is widely distributed in plant tissues. It is formed in the hydrolysis of starch, and its relationships in the plant to starch and to other sugars will be considered later. It is a white crystalline substance soluble in water and alcohol. In constitution it is a glucose-α-glucoside:

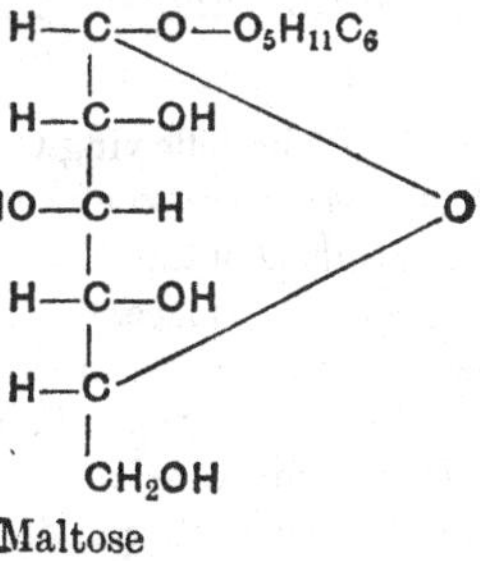

Maltose

It reduces Fehling's solution; but less readily than glucose. With phenylhydrazine hydrochloride and sodium acetate it forms an osazone (m.p. 206° C.), which is more soluble than glucosazone and crystallizes in broader flatter needles. Maltose is dextro-rotatory.

Expt. 45. *Tests for maltose.* The tests *a*, *b*, *c* and *e*, should be performed with a 0·2% solution of maltose ; test *d* with a 2 % solution (see also glucose, p. 50).

(*a*) *Moore's test.* A positive reaction is given.

(*b*) *Trommer's test.* A positive reaction is given.

(*c*) *Fehling's test.* Reduction takes place, but less strongly than with glucose.

(*d*) *Osazone test.* Take 10 c.c. of the solution and treat as for glucosazone. The crystals of maltosazone will be found to be much broader than those of glucosazone.

(*e*) *Hydrolysis.* Take 20 c.c. of the sugar solution and add 2 c.c. of concentrated hydrochloric acid. Heat in a boiling water-bath for half an hour. Neutralize and test for the osazone. Glucosazone will be formed.

Sucrose. Sucrose or cane-sugar is very widely distributed in plants, in leaves, stems, roots, fruits, etc. It is a white substance which crystallizes well, and is soluble in water and alcohol. As previously stated it is hydrolyzed by dilute acids and by invertase into one molecule of glucose and one molecule of laevulose. It is formed by the condensation of glucose and laevulose with the elimination of water. Its constitution is in all probability as follows:

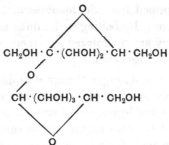

so that both the ketone and aldehyde groups are rendered inactive. It does not reduce Fehling's solution and does not form an osazone. It is dextro-rotatory.

Expt. 46. *Tests for cane-sugar.* The following tests should be made with a 1 % solution of pure crystalline cane-sugar (see also glucose, p. 50).

(*a*) *Moore's test.* A negative result is obtained.

(*b*) *Fehling's test.* No reduction takes place.

(*c*) *a-Naphthol test.* A positive result is given since sucrose yields laevulose.

(*d*) *Hydrolysis.* To a few c.c. of the solution add a drop of strong sulphuric acid and boil for two minutes. Then neutralize with caustic soda using litmus as indicator. Boil again and add Fehling's solution drop by drop. A reduction takes place owing to the inversion of the cane-sugar by sulphuric acid.

(*e*) *Seliwanoff's test.* A positive result is obtained owing to the liberation of laevulose.

TRI- AND TETRASACCHARIDES.

Several trisaccharides, condensed from various hexoses or pentoses are known. **Raffinose** (fructose, glucose and galactose) has been isolated from the seed of the Cotton Plant (*Gossypium*), from the Beet (*Beta*) and other plants. **Rhamninose** (galactose and two molecules of rhamnose) occurs in the fruit of *Rhamnus infectoria*. **Gentianose** (fructose and two molecules of glucose) has been isolated from the root of Gentian (*Gentiana*). **Melicitose** (fructose and two molecules of glucose) occurs in a manna[1] which exudes from the twigs of the Larch (*Larix*) and Douglas Fir (*Pseudotsuga*).

A tetrasaccharide, **stachyose** (fructose, glucose and two molecules of galactose) has been isolated from tubers of *Stachys tubifera*, from White Jasmine (*Jasminum*) and other plants.

POLYSACCHARIDES.

These substances are formed by condensation, with elimination of water, from more than three molecules of monosaccharides.

PENTOSANS.

It has already been mentioned that condensation products of the pentoses, the pentosans, are widely distributed. The two most frequently occurring pentosans are xylan and araban. No enzymes are known which hydrolyze the pentosans. It is characteristic of xylan and araban that they form copper compounds in Fehling's solution in presence of excess of alkali.

Araban. This pentosan may be regarded as a condensation product of arabinose as already indicated. It occurs in various gums (Gum Arabic, Cherry Gum) frequently in combination with other substances. On hydrolysis with acids, araban yields arabinose. (See also gums and arabinose.)

Expt. 47. *Preparation of araban from Gum Arabic.* (Salkowski, 30.) Weigh out 20 gms. of gum arabic and dissolve in 500 c.c. of warm water in a large evaporating dish on a water-bath. Then add 200 c.c. of Fehling's solution and excess of strong caustic soda solution. The araban will be precipitated as a white gummy mass which will settle at the bottom of the dish. Filter off through muslin. Take up the precipitate in the minimum quantity of dilute hydrochloric acid (1 pt. of acid : 1 pt. of water), and then add alcohol. The araban separates out as a white precipitate. Wash away the copper chloride with alcohol.

Expt. 48. *Hydrolysis of araban.* The araban from the last experiment is put into a round-bottomed flask with about 200 c.c. of 2 % sulphuric acid and heated on a water-bath for 2 hours, the flask being fitted with an air condenser (see p. 46).

---

[1] Manna is a name given to exudations from the branches of various trees and shrubs. Sometimes the flow is assisted artificially as in the case of the Manna Ash (*Fraxinus Ornus*) where the product, consisting almost entirely of the polyhydric alcohol, mannitol, is of commercial value as a drug, etc. In other cases, the manna exudes as the result of the attacks of insects. Mannas appear to be readily soluble in water to clear, non-sticky solutions, thereby differing from gums and resins.

Then neutralize the liquid with calcium carbonate, filter from calcium sulphate, and concentrate on a water-bath. Some of the solution of arabinose should be tested with all the tests given in Expt. 39. The sugar can be extracted from the syrup with 90% alcohol, but it crystallizes only with difficulty.

Xylan. This pentosan occurs in lignified cell-walls, and is the chief constituent of " wood gum." It is found in the wood of many trees (not Coniferae), in bran, in wheat and oat straw, in maize cobs, in the shells of coconuts and walnuts, in the testa of the cotton (*Gossypium*) and in many other tissues: also in some gums. On hydrolysis, xylan yields xylose; hence wood shavings, bran, straw, etc., will give the pentose reactions on hydrolysis.

Expt. 49. Detection of pentose from pentosans in bran, sawdust and straw. Take a small quantity of bran and boil it up several times with 98% alcohol, filtering off the alcohol after each treatment. This should remove any sugars or glucosides present. Allow the alcohol to evaporate off from the bran, and then make the following tests for pentoses (see Expt. 39):

(a) Heat, for about one minute, a small quantity of the bran in a test-tube, with sufficient concentrated hydrochloric acid to cover it. Care should be taken not to char the material. Then add as much solid orcinol as will lie on the tip of a penknife. Heat gently again for a few seconds. Then add one or two drops of strong ferric chloride solution; a green coloration will be produced. Add amyl alcohol and the green colour will pass into the alcohol.

(b) Heat again another portion of the bran with the same quantity of concentrated hydrochloric acid in a test-tube, but this time heat more strongly. After heating a few minutes place a piece of filter-paper soaked in a solution of aniline acetate in the mouth of the test-tube. A cherry-red coloration will denote the formation of furfural.

The above method and tests with bran may be repeated in exactly the same way using sawdust or straw.

Expt. 50. Preparation of xylan from sawdust. Extract one kilo of sawdust with 4 litres of 1–2% ammonia solution for 24 hrs. Then filter off the ammoniacal solution through muslin and repeat the extraction. The xylan is insoluble in ammoniacal solution, and in this way colouring matters are removed. Finally wash the sawdust well with water and press dry from the liquid. Then add to the sawdust sufficient 5% caustic soda solution to make a thick mush (about 1000–1500 c.c.) and allow it to stand for 24 hrs. in a warm place. The alkaline solution is then pressed out through calico and filtered through filter-paper. To the clear filtrate add an equal volume of 96% alcohol which will precipitate the xylan as a sodium compound. Filter off this precipitate, wash with alcohol, and decompose with alcohol to which a little strong hydrochloric acid has been added to remove the sodium. The free xylan is again washed with alcohol, and can be dried by washing with absolute alcohol and ether and finally in a desiccator. It is a dirty-white powder which is almost insoluble in water. Make the tests for pentoses (see Expt. 39) on a little of the solid xylan. The reaction will be given in each case.

Expt. 51. *Hydrolysis of xylan.* Put the xylan obtained in the last experiment in a round-bottomed flask fitted with an air condenser (see p. 46). Add 100 c.c. of $4\,^0/_0$ sulphuric acid and heat on a water-bath for 4 hrs. Neutralize the solution with calcium carbonate, filter from calcium sulphate and concentrate on a water-bath. Test a portion for pentoses (see Expt. 39) and a positive reaction will be obtained. To a small quantity add also a few drops of Fehling's solution and boil. Reduction will take place.

To the remainder of the xylose solution add bromine (see p. 47) gradually until there is excess. Then remove the excess of bromine by warming on a water-bath. Neutralize the solution, which contains xylonic acid, with cadmium carbonate and evaporate on a water-bath. Extract the residue with alcohol and filter. On concentrating the alcoholic extract, white prismatic needles of cadmium xylonate separate out.

It has been shown that pentosans, xylan and probably araban, occur in leaves (Davis, Daish and Sawyer, 17). It is likely that the xylan is widely distributed in all tissues since it forms a constituent of lignified cell-walls. (See Appendix, p. 187.)

Expt. 52. *Detection of pentoses from pentosans in leaves.* (Davis, Daish and Sawyer, 17.) Take two large leaves of the Sunflower (*Helianthus annuus*). Tear into small pieces and drop into boiling $98\,^0/_0$ alcohol in a flask. Boil well and filter off the alcohol. Repeat until all the green colour is removed. Then dry off the alcohol and grind up the leaf residue. Perform the test for pentoses (Expt. 39 *a* and *c*) on the dry leaf tissue. It should give the above tests showing the presence of pentosans.

Leaves of the Violet (*Viola odorata*) and Nasturtium (*Tropaeolum majus*) may also be used.

Expt. 53. *Method for determination of pentosans in tissues, bran and leaves, etc.* Weigh out 2 gms. of bran, put it into a round-bottomed flask, add 100 c.c. of $12\,^0/_0$ hydrochloric acid and fit the flask with a water condenser. Heat gently over wire gauze and distil into a solution of phloroglucinol in $12\,^0/_0$ hydrochloric acid. A green precipitate of furfural phloroglucide is formed which eventually becomes almost black. For accurate estimations of pentosans this is filtered off and weighed on a Gooch crucible. The same method may be used with leaf residue prepared as in Expt. 52.

STARCHES.

Starch. This is a very widely distributed substance in plants. It occurs as solid grains throughout the tissues, in leaves, stems, roots, fruits and seeds. It is absent, however, from a number of Monocotyledons, e.g. *Iris*, Snowdrop (*Galanthus*), *Hyacinthus*, etc. (Blackman, 5). It forms one of the chief reserve materials of plants, that is, it is synthesized from sugar when carbon assimilation and carbohydrate synthesis are in progress, and is stored in the solid form in tissues as grains. In other circumstances of the plant's existence, when material for metabolism is not available from carbon assimilation, as for instance in germinating seeds or growing bulbs or rhizomes, the starch is hydrolyzed into dextrin and soluble sugar, which is translocated and used as a basis for metabolism. During the night in leaves there is also a similar hydrolysis of the starch which has been temporarily stored from the excess of sugar synthesized during the day.

Starch has a very large molecule and thus a high molecular weight. It is insoluble in cold water. When heated with a little water it gives starch paste, but on boiling with water it gives an opalescent "solution" which really contains starch in the colloidal state as an emulsoid. In this condition it does not diffuse through dialyzing membranes and does not depress the freezing point of water. The "solution" cannot, strictly speaking, be filtered, but generally, when hot, it passes to some extent through ordinary filter-paper. Starch is insoluble in alcohol and is precipitated by it.

The most characteristic reaction of starch is the blue colour it gives with iodine solution. This blue colour disappears on heating, but reappears again on cooling. Starch is precipitated from "solution" by half saturation with ammonium sulphate: it does not reduce Fehling's solution.

By boiling with dilute acids, starch is first converted into "soluble starch" which still gives a blue colour with iodine. On further boiling, various dextrins (see dextrins) are obtained which give either purple, red or no colour with iodine. The final product, after prolonged boiling with acids, is glucose. Hydrolysis with diastase yields dextrin and maltose (see diastase, p. 75).

Expt. 54. *Preparation of starch from Wheat.* Starch may be prepared from a cereal by the following method.

Take 25 gms. of flour and make it up into a dough with a little water. Allow it to stand for half an hour. Then tie a piece of muslin over the top of a beaker which is filled with water. Place the dough on the top of the muslin and rub it gently with a glass rod. The starch will be separated from the gluten, and will be washed through the muslin and on standing will sink to the bottom of the beaker. Allow this to stand till the starch has settled, then decant off the bulk of the liquid. Filter off the starch, and wash well with water, then with alcohol and finally with ether. Dry in the steam-oven.

For the detection of starch in green leaves, see Expt. 77.

Expt. 55. *Tests for starch.* Take a small quantity of the starch prepared in the previous experiment (or use commercial potato starch) and shake up with a little cold water in a test-tube. Filter, and test the filtrate with a drop of iodine (in potassium iodide) solution. No blue colour is obtained. Pour a drop of the iodine solution on the residue in the filter. It turns deep blue.

Weigh out 2 gms. of the starch prepared in the last experiment, and mix it into a thin cream with a little water. Boil rather more than 100 c.c. of water in an evaporating dish, and then gradually add to it the starch paste, keeping the water boiling all the time. An opalescent "solution" is obtained. With a few c.c. of the solution in each case make the following tests:

(a) Add 1–2 drops of iodine solution. A blue colour is obtained. Heat the solution: the blue colour disappears, but reappears on cooling.

(b) Add an equal volume of alcohol: the starch is precipitated.

(c) Add an equal volume of saturated ammonium sulphate solution : the starch is precipitated, i.e. by half saturation with this salt.

(d) Add basic lead acetate solution: the starch is precipitated.

Expt. 56. *Hydrolysis of starch.* To 50 c.c. of the starch solution prepared in the last experiment add 1 c.c. of strong sulphuric acid. Boil for 10–20 minutes in a round-bottomed flask. Test a portion of the solution with iodine from time to time; a purple, red or brown colour is formed due to the dextrin produced in hydrolysis. To the remainder of the solution after neutralization, using litmus as indicator, add some Fehling's solution and boil. Reduction takes place owing to the glucose formed in hydrolysis.

DEXTRINS.

These compounds occur in the plant as transitory substances, since they are formed as intermediate products of the hydrolysis of starch by diastase. They are also formed on heating starch or by boiling it with mineral acids (see previous experiment). The hydrolysis of starch to dextrins is fairly rapid, but the conversion of dextrins into maltose is a much slower process.

Both starch and dextrins have the same empirical formula. Various forms of the latter have been identified, such as amylodextrin which gives a blue colour with iodine, erythrodextrin which gives a brownish-red colour with iodine, and achroodextrin which gives no colour with iodine. The dextrins are readily soluble in water; they are precipitated by alcohol but not by basic lead acetate. On hydrolysis with acids, they are converted into glucose.

Expt. 57. *Preparation of dextrin by hydrolysis of starch.* (a) *By diastase from leaves of the Pea* (Pisum sativum). Weigh out 10 gms. of commercial potato starch and make it into a solution in 250 c.c. of boiling distilled water as in Expt. 55 and cool. Then weigh out 10–15 gms. of fresh leaflets of the Pea (*Pisum sativum*) and pound them well in a mortar. Add to the pounded mass 100 c.c. of distilled water and a few drops of chloroform (see maltase, p. 77) and filter. The filtrate will contain diastase (see also Expt. 78). Then add the diastase extract to the starch solution in a flask, plug with cotton-wool and put in an incubator for 48 hrs. If a little of the liquid is withdrawn from time to time and tested with iodine, it will be found that the blue colour due to the starch gradually disappears and is replaced by the brownish-red colour due to dextrin. After 48 hrs. there will be no trace of blue colour; then filter the liquid and concentrate the filtrate on a water-bath to a syrup. Treat the residue with about 30 c.c. of 96–98 % alcohol and filter. A sticky mass of dextrin is left which should be extracted with a little hot alcohol and then reserved for the next experiment. To show the presence of maltose, the combined alcoholic extracts are evaporated to dryness on a water-bath, the residue taken up in a little water and the osazone test made (see p. 50) with the solution. Crystals of maltosazone will separate out.

(b) *By diastase from germinating Barley* (Hordeum vulgare). Weigh out about 25 gms. of barley grains and allow them to germinate by soaking and spreading on damp blotting-paper for 5–7 days. Pound the grains well in a mortar, add 100 c.c. of water, allow to stand for 2–3 hrs. and filter. Precipitate the filtrate with alcohol and allow to stand for 24 hrs. Filter off the precipitate, take up in water and add it to the barley starch "solution," together with a few drops of chloroform. Proceed as with (a) only the time for hydrolysis may be much shorter, i.e. 6–12 hrs.

Expt. 58. *Tests for dextrin.* Make a solution of the dextrin prepared in the last experiment (or use commercial dextrin) and note that it is very soluble in water. With the solution make the following tests:

(a) Add a little iodine solution. If erythrodextrin is present, a reddish-brown colour is produced. Heat the solution and the colour will disappear. Cool again and the colour will reappear. If only achroodextrin is present, no colour will be given with iodine.

(b) Add an equal volume of strong alcohol. The dextrin is precipitated.

(c) Add an equal volume of saturated ammonium sulphate solution, i.e. half saturation with ammonium sulphate. The dextrin is not precipitated.

(d) Add some basic lead acetate solution: the dextrin is not precipitated.

INULIN.

Inulin. This substance occurs as a soluble "reserve material" in the cell-sap of the underground stems, roots and also leaves of a number of plants, especially members of the Compositae, e.g. Dahlia (*Dahlia variabilis*), Jerusalem Artichoke (*Helianthus tuberosus*), Chicory (*Cichorium Intybus*) and the Dandelion (*Taraxacum officinale*). It is said to occur also in the Campanulaceae, Lobeliaceae, Goodeniaceae, Violaceae and many Monocotyledons (*Hyacinthus, Iris, Muscari* and *Scilla*).

Inulin is a condensation product of laevulose to which it bears much the same relation as starch to glucose. It is a white substance, soluble in water and insoluble in alcohol. It crystallizes out in the cells, in which it occurs, in characteristic sphaero-crystals on addition of alcohol to the tissues. It is hydrolyzed by mineral acids to laevulose: also by the enzyme inulase which occurs in the plant.

Expt. 59. *Extraction of inulin.* Cut off the tubers from two Dahlia (*Dahlia variabilis*) plants, wash well, and put them through a mincing machine. Carefully collect the liquid and the crushed tuber, and boil well with sufficient water to cover the crushed material. Add also some precipitated calcium carbonate to neutralize any free acids present. Then filter through fine muslin, and to the filtrate, which should again be made quite hot, add lead acetate solution until a precipitate (of mucilaginous substances, etc.) ceases to be formed. Care should be taken to avoid the addition of a large excess of lead acetate. Filter off the lead precipitate, and saturate the filtrate with sulphuretted hydrogen till all excess lead is removed. Filter off the lead sulphide, neutralize the filtrate to phenolphthalein with ammonia, and evaporate to half bulk or less on a water-bath, when the inulin will probably begin to deposit. Then pour into an equal volume of alcohol, and allow to stand for one or two days. The crude precipitate of inulin is filtered off, dissolved in a small amount of water, and reprecipitated with alcohol. It can be washed with alcohol and ether and dried over sulphuric acid.

The Artichoke (*Helianthus tuberosus*) may also be used, about 12 tubers being necessary.

Expt. 60. *Tests for inulin.* Make a solution of some of the inulin prepared in Expt. 59 in hot water. It will readily dissolve giving a clear solution. With the solution make the following tests :

(*a*) Make a very dilute solution of iodine and add to it a drop or two of inulin solution : the brown colour will be unaffected.

(*b*) Boil some inulin solution with a little Fehling: no reduction takes place.

If the inulin solution which is being used should reduce Fehling it indicates that sugar is present as impurity. If this is the case, then a little of the solid inulin should be washed free from sugar by means of alcohol before proceeding with the following test.

(*c*) To 5 c.c. of Seliwanoff's solution add a few drops of inulin solution and boil. A red coloration is formed. This reaction is also due to the presence of laevulose (see laevulose, p. 53).

Expt. 61. *Hydrolysis of inulin.* Some inulin is dissolved in very dilute hydrochloric acid (about 0·5 %) and heated on a water-bath for half an hour in a round-bottomed flask provided with an air condenser (see p. 46). The solution is then neutralized with sodium carbonate and concentrated on a water-bath. With the concentrated solution make the following tests :

(*a*) Boil with a little Fehling: the solution is rapidly reduced.

(*b*) Make the osazone test (see p. 50). Glucosazone crystals will be found to be formed on microscopic examination. (Laevulose forms the same osazone as glucose.)

(*c*) Make the test (*c*) of the last experiment. A positive result will be given.

MANNANS.

The mannans which have already been mentioned (see p. 52) are condensation products of the hexose, mannose. They occur most frequently, either mixed, or in combination, with the condensation products of other hexoses and pentoses (glucose, galactose, fructose and arabinose) as galactomannans, glucomannans, fructomannans, mannocelluloses, etc. Such mixtures or compounds of which mannans form a constituent are widely distributed in the seeds of many plants, i.e. Palms (including the Date-palm), Asparagus (*Ruscus*), Clover (*Trifolium*), Coffee Bean (*Coffea arabica*), Onion (*Allium Cepa*) and of members of the Leguminosae, Rubiaceae, Coniferae and Umbelliferae. In seeds the mannans may constitute, together with cellulose, the thickened cell-walls of the endosperm and are included in the term "reserve- or hemi-cellulose" though they are not strictly celluloses. "Vegetable ivory," which is the endosperm of the Palm, *Phytelephas macrocarpa*, contains considerable quantities of a mannan and is used as a source of mannose. Mannans, in addition, form constituents of certain mucilages, as for instance those in Lily bulbs (*Lilium candidum, L. bulbiferum, L. Martagon* and others) (Parkin, 25) and tubers of various genera of the Orchidaceae : they are also found in the roots of the Dandelion (*Taraxacum*), *Helianthus* and Chicory, *Asparagus* and Clover, and in the wood and leaves of various trees.

Many of the mannans, unlike true celluloses, are readily hydrolyzed by dilute hydrochloric and sulphuric acids. The mannan in the Coffee Bean, however, is hydrolyzed with difficulty.

GALACTANS.

These substances bear the same relationship to the hexose, galactose as the mannans to mannose, that is, they are condensation products of galactose (see p. 51). Similarly they frequently occur, together with the condensation products of other sugars, as galactoaraban, galactoxylan, galactomannan, etc. As such they form constituents of many gums and mucilages and of the cell-walls of the reserve tissue of seeds, i.e. the Coffee Bean (*Coffea arabica*), the Bean (*Faba*), the Lupin (*Lupinus*), the Paeony (*Paeonia*), the Kidney Bean (*Phaseolus*), the Date (*Phoenix*), the Pea (*Pisum*), the Nasturtium (*Tropaeolum*) and many others (Schulze, Steiger and Maxwell, 32).

GUMS.

These substances occur widely distributed among plants, especially trees. Some gums are wholly soluble in water giving sticky colloidal solutions : others are only partially soluble. They are all insoluble in alcohol. In the solid state they are translucent and amorphous.

Chemically the gums are varied in nature ; they may in general be regarded as consisting of complex acids in combination with condensation products of various sugars, such as araban, xylan, galactan, etc. On hydrolysis they give mixtures of the corresponding sugars, arabinose, xylose, galactose, etc., in varying proportions, though in some cases one sugar preponderates.

Some of the best-known gums are the following :

Gum Arabic (arabin). This substance is obtained from an Acacia (*Acacia Senegal*), a native of the Soudan. The gum exudes from the branches. Other species of Acacia yield inferior gums. Gum arabic is a mixture of the calcium, magnesium and potassium salts of arabic acid, a weak acid of which the constitution is unknown, in combination with araban and galactan.

Gum Tragacanth. This is a product from several Tragacanth shrubs which are species of *Astragalus* (Leguminosae), chiefly *A. gummifer*. It is obtained by wounding the stem and allowing the gum to exude and harden. On hydrolysis it gives a mixture of complex acids and various sugars such as arabinose, galactose and xylose.

Cherry Gum (cerasin) occurs in the wood of the stems and branches of the Cherry (*Prunus Cerasus*), the Bird Cherry (*P. Padus*), the Plum (*P. domestica*), the Almond (*P. Amygdalus*) and other trees of the Rosaceae. It exudes from fissures of the bark. On hydrolysis it yields almost entirely arabinose.

Expt. 62. *Reactions of Gum Arabic.* Put a little gum arabic into an evaporating dish and add a little water. Heat gently and stir. The gum will slowly dissolve, giving a thick sticky solution which does not solidify or gel on cooling. Make the following tests, using a little of the gum solution in a test-tube each time.

(*a*) Add a little alcohol. The gum is precipitated.

(*b*) Add a little Fehling's solution and boil. No reduction takes place.

The three following experiments show the presence of pentosan complexes in the gum (see also Expt. 39, p. 45):

(*c*) Add a little phloroglucinol to the gum and then strong hydrochloric acid. No colour is produced. Now heat, and a cherry-red colour appears.

(*d*) Heat the gum solution with a little concentrated hydrochloric acid and then add a trace of orcinol. Warm again and then add one or two drops of strong ferric chloride solution. A green coloration will be produced.

(*e*) Heat the gum solution strongly with hydrochloric acid, and, after heating for a few minutes, place a piece of filter-paper soaked in a solution of aniline acetate in the mouth of the test-tube. A cherry-red coloration indicative of furfural will be formed.

Expt. 63. *Hydrolysis of Gum Arabic.* Weigh out 10 gms. of gum arabic. Put it into a round-bottomed flask and add 100 c.c. of water and 4 c.c. of strong sulphuric acid. Warm gently until the gum goes into solution. Then fit the flask with an air condenser (see p. 46) and heat on a water-bath for about 4 hrs. Cool the solution, and neutralize with barium carbonate. Filter off the barium sulphate and concentrate the solution on a water-bath. Boil a drop or two of the syrup with Fehling's solution and show that reduction takes place. (The original gum either does not reduce Fehling at all, or, if so, only slightly.) Then add a little nitric acid (sp. gr. 1·15, see

Expt. 43) to the syrup and heat on a water-bath almost to dryness. Pour the residue into about 100 c.c. of water and allow to stand. A microcrystalline precipitate of mucic acid is formed showing the presence of galactose (see p. 52) as a product of hydrolysis.

MUCILAGES.

The characteristic of these substances is that they swell up in water and produce colloidal solutions which are slimy.

Mucilages are widely distributed and may occur in any organ of the plant. Sometimes they are confined to certain cells, mucilage sacs or canals. They are distinguished from the pectic substances by the fact that they do not gelatinize. Some of the best known examples of mucilage-containing tissues are those in the root and flower of the Hollyhock (*Althaea rosea*): in succulent plants (*Aloe, Euphorbia*), in bulbs (*Scilla, Allium*) and tubers (*Orchis Morio*): in seeds of Flax or Linseed (*Linum*) and in fruits of Mistletoe (*Viscum album*).

The mucilages vary in composition. They appear to be largely, if not wholly, condensation products of various sugars (galactose, mannose, glucose, xylose, arabinose), similar constituents to those of many gums and hemicelluloses. On hydrolysis various mixtures of sugars are produced. Of the mucilages, that from linseed has been thoroughly investigated. It has been found on hydrolysis to give sugars only, e.g. arabinose, xylose, glucose and galactose. In this respect mucilages differ from gums, since the latter have always some other accompanying substance in addition to sugars.

Expt. 64. *Preparation and properties of mucilage from Linseed* (Linum) (Neville, 23). Take about 60 gms. of linseed and let it soak for 24 hrs. in 300 c.c. of water. Then separate the slime from the seeds by squeezing through muslin, and add to the liquid about twice its volume of 96–98 % alcohol. The mucilage is precipitated as a thick slimy precipitate. Filter off the precipitate and wash with alcohol. By washing with absolute alcohol and ether and finally drying in a desiccator, the mucilage may be obtained as a powder.

. Add water to some of the mucilage. It swells up and finally gives an opalescent solution. Make with it the following tests:

(*a*) Add iodine. No colour is given.

(*b*) Add a little Fehling's solution and boil. No reduction takes place.

Expt. 65. *Hydrolysis of Linseed mucilage.* Put the remainder of the mucilage in a round-bottomed flask and add 50 c.c. of 4 % sulphuric acid. Fit the flask with an air condenser (see p. 46) and heat for at least four hours on a water-bath. Cool and neutralize with barium carbonate. Filter off the barium sulphate, and concentrate the filtrate on a water-bath. With the concentrated solution make the following tests:

(a) Add a few drops to a little boiling Fehling solution. Reduction immediately takes place.

(b) Make the phloroglucinol, orcinol and furfural tests for pentoses, using a small quantity only of the hydrolysis mixture for the tests. A positive result will be given in each case. The pentoses, arabinose and xylose, are responsible for these reactions.

(c) Add to some of the solution phenylhydrazine hydrochloride, sodium acetate and a little acetic acid, and leave in boiling water for half an hour for the osazone test [see Expt. 41 (d)]. A mixture of osazones will separate out, among which glucosazone can be identified.

(d) Concentrate the remainder of the solution and then add some nitric acid of sp. gr. 1·15 (see Expt. 43). Evaporate down on a water-bath to one-third of the bulk of the liquid and then pour into about 100 c.c. of water. A white microcrystalline precipitate of mucic acid will separate out, either at once or in the course of a day or two. This demonstrates the presence of galactose.

PECTIC SUBSTANCES.

These substances are considered at this point since they are said to be present in intimate connexion with cellulose in the cell-wall, especially the middle lamella, of many tissues. The pectic substances are frequently found in the juices of succulent fruits in which the tissues have disintegrated, such as red currants and gooseberries. They have been isolated chiefly from fleshy roots, stems and fruits, as, for instance, from turnips, beetroot, rhubarb stems, oranges, apples, cherries and strawberries; more recently, also, from cabbage, onions and pea-pods. Recent investigations point to the fact that all these tissues contain the same pectic material, and it is possible that all such substances may be identical. (See Appendix, p. 185.)

The chief pectic compound occurring in the cell-wall, probably in combination with cellulose, is now known as pectin, though it was provisionally termed pectinogen (Schryver and Haynes, 31). It is extracted, in the form of the ammonium salt, by treating the tissue residue (after expressing the juice) with warm dilute ammonium oxalate solution. From this solution, either the salt, or, after acidification, pectin itself, can be precipitated as a very bulky gelatinous mass by adding alcohol. Pectin is an acid and is soluble in water, giving a thick opalescent solution; its sodium, potassium, ammonium and calcium salts are also soluble. Pectin solution, therefore, is not precipitated either by acid or by dilute solutions of calcium salts.

In the case of juicy fruits, such as currants and gooseberries, the pectin can be precipitated as a gelatinous precipitate by adding alcohol to the expressed juice. In the case of fleshy fruits, stems and roots, the juice, as a rule, contains but little pectin and the procedure is as follows. The tissues are thoroughly disintegrated in a mincing machine and pressed free from all juice in a powerful press. The residue is then dried, finely ground, washed with water and finally extracted with dilute ammonium oxalate solution in which pectin is soluble. The extract is concentrated and the pectin precipitated by alcohol. It may be purified by reprecipitation. When dried it forms an almost colourless granular powder. Put into water it absorbs large quantities of liquid and dissolves slowly, giving an opalescent solution with a distinctly acid reaction.

When pectin solution is treated with normal caustic soda at ordinary temperatures, the sodium salt is first formed and this is rapidly changed into the salt of another substance termed pectic (cytopectic) acid (Clayson, Norris and Schryver, 7). Pectic acid is insoluble in water and is readily converted into a gel under certain conditions; its calcium salt is also insoluble. Thus, if a solution of pectin, made alkaline with caustic soda, is allowed to stand for ten minutes, on adding acid a gelatinous precipitate of pectic acid is formed, and on adding calcium chloride solution, a gelatinous precipitate of the calcium salt of pectic acid. A similar precipitate is also formed when lime water is added in excess to a solution of pectin and it is allowed to stand.

If the tissue residue is first treated with caustic soda solution, the pectin is changed *in situ* into the pectic acid which, though not itself extracted with caustic soda, can be subsequently extracted by ammonium oxalate solution and separated as a gel by addition of acid.

Analyses of pectic acid from various sources, i.e. apples, oranges, strawberries, cabbage, onions, pea-pods, rhubarb and turnips, have led to various suggestions as to its formula. There is also evidence that it contains one pentose group. This can be detected and estimated by the furfural phloroglucide method (see Expt. 53 and also Appendix, p. 187).

Expt. 66. *Extraction and reactions of pectin.* Take about half a pound of red currants and squeeze out the juice through fine muslin into a large beaker. Then add to the juice about 2–3 times its bulk of 96–98 % alcohol. A bulky gelatinous precipitate of pectin will separate out. Allow the precipitate to stand for a time in the alcohol and then filter off. Wash with alcohol and finally press free from liquid. Dissolve the precipitate in as little water as will enable it to go into solution. To two small portions of the solution add respectively (a) a few drops of strong hydrochloric acid, (b) an excess of 5 % calcium chloride solution. Note that no precipitate is formed in either case.

Expt. 67. *Conversion of pectin into pectic acid, and reactions of pectic acid.* Take about one-third of the pectin solution prepared in Expt. 66, make it alkaline with 4 % caustic soda, and let it stand for about 10–15 minutes. Then divide the solution into two parts and add respectively (a) sufficient strong hydrochloric acid to acidify,

(b) excess of 5 % calcium chloride solution. In the first case a gel of pectic acid is formed: in the second case a gelatinous precipitate of the calcium salt of pectic acid.

To a further quantity of the pectin add excess of lime water and let it stand. The gelatinous calcium precipitate will separate out in a short time.

Expt. 68. *Detection of the pentose group in pectin.* Filter off the pectic acid gel obtained in the last experiment and allow it to dry. Then test for the pentose group by the orcinol, phloroglucinol and furfural tests (see Expt. 39). All results will be found to be positive.

The extraction of pectin, etc. in the above experiments can equally well be carried out with other material, e.g. ripe gooseberries, raspberries and strawberries, using exactly the same methods.

Expt. 69. *Preparation of pectin from Turnips.* Take two full-sized turnips and mince them *finely* in a mincing machine. Then wrap the mass in a piece of strong unbleached calico and press out the juice as completely as possible in a press. The juice contains little pectin and can be thrown away. The pressed mass is then thrown into about 200 c.c. of freshly prepared 0·5 % ammonium oxalate solution heated to 80–90° C. on a water-bath and stirred to make a paste. The liquid is again rapidly pressed out in the press. To the viscid extract an equal volume of 96 % alcohol is added, and the ammonium salt of pectin separates out as a voluminous gelatinous precipitate. This is filtered off and, when pressed free from alcohol and dried, can be used for tests as in the previous experiments.

The gelatinization of pectin can also be brought about by certain enzymes termed pectases which are found in the juices of various plants, i.e. root of Carrot (*Daucus Carota*) and leaves of Lucerne (*Medicago sativa*), Lilac (*Syringa vulgaris*) and Clover (*Trifolium pratense*).

Expt. 70. *Action of pectase on pectin.* Make an extract of either Lucerne or Clover leaves by pounding them in a mortar with a little water, and then filter. Add the filtrate to some of the pectin solution prepared in Expt. 66 or 69. On standing a gelatinous precipitate will be produced. Should the reaction be slow, it may be accelerated by placing the mixture in an incubator.

CELLULOSES.

Celluloses are very important polysaccharides. They form constituents of the structural part of all the higher plants. The cell-wall of the young cell consists entirely of cellulose, but in older cells the walls may be lignified, cuticularized, etc., i.e. the cellulose may be accompanied by other substances such as lignin, cutin, mucilage, etc. In the light of these facts the term cellulose is made to include:

1. Normal celluloses.
2. Compound celluloses:
    (a) Ligno-celluloses.
    (b) Pecto-celluloses.
    (c) Adipo- or cuto-celluloses.
3. Pseudo- or Reserve-celluloses.

True or normal cellulose. Of this substance, as we have said, many cell-walls are composed. The most familiar form of cellulose is cotton, which consists of hairs, each being a very long empty cell, from the testa or coat of the seed of the Cotton plant (*Gossypium herbaceum*). Crude cotton (i.e. the hair cell-walls) is not quite pure cellulose, but contains a small amount of impurity from which it is freed by treatment first with alkali and subsequently with bromine or chlorine. All kinds of cotton material, cotton-wool, and the better forms of paper (including filter-paper) may be regarded as almost pure cellulose.

Pure cellulose is a white, somewhat hygroscopic, substance. It is insoluble in water and all the usual solvents for organic substances. It is, however, soluble in a solution of zinc chloride in hydrochloric acid in the cold, and in a solution of zinc chloride alone on warming. It is also soluble in ammoniacal cupric oxide (Schweizer's reagent).

In addition cellulose is soluble in concentrated sulphuric acid, which on standing converts it first into a hydrate and then finally into glucose. If, however, water is added to the sulphuric acid solution as soon as it is made, the gelatinous hydrate of cellulose is precipitated. This substance is termed "amyloid" since it gives a blue colour with iodine. Concentrated nitric acid converts cellulose into nitrates, of which one is the substance, gun-cotton. In 10 % alkalies cotton fibres thicken and become more cylindrical. This procedure has been employed by Mercer to give a silky gloss to cotton, and the resultant product is called mercerized cotton.

Expt. 71. *The colour tests and solubilities of cellulose.*

(*a*) Dip a little cotton-wool into a solution of iodine in potassium iodide. Then put the stained wool into an evaporating dish and add a drop or two of concentrated sulphuric acid. A blue coloration is given. This is due to the formation of the hydrate "amyloid" mentioned above.

(*b*) Dip some cotton-wool into a calcium chloride iodine solution. (To 10 c.c. of a saturated solution of calcium chloride add 0·5 gm. of potassium iodide and 0·1 gm. of iodine. Warm gently and filter through glass-wool.) A rose-red coloration is produced which eventually turns violet.

(*c*) Heat a strong solution of zinc chloride (6 pts. of zinc chloride to 10 pts. of water) in an evaporating dish and add 1 part of cotton-wool. The cellulose will in time become gelatinized, and if a little water is added from time to time, a solution will eventually be obtained on continuous heating.

(*d*) Make a solution of zinc chloride in twice its weight of concentrated hydrochloric acid and add some cotton-wool. The wool will rapidly go into solution in the cold.

(*e*) Add some cotton-wool to an ammoniacal copper oxide solution and note that it dissolves. (To a strong solution of copper sulphate add some ammonium chloride and then excess of caustic soda. Filter off the blue precipitate of cupric hydroxide, wash well, dry thoroughly, and dissolve in strong ammonia.) Add strong hydrochloric acid and the cellulose is precipitated out again. Then add water and wash the precipitate until it is colourless. Test the roughly dried precipitate with a little iodine and strong sulphuric acid. A blue coloration is given.

All the above tests may be repeated with threads from white cotton material, with filter-paper and good white writing paper.

Try tests (a) and (b) with newspaper, and note that they are not so distinct as with writing paper owing to the presence of ligno-cellulose (see Expt. 73).

Expt. 72. *Hydrolysis of cellulose by acid.* Dissolve as much filter-paper as possible in 5 c.c. of concentrated sulphuric acid and when all is in solution pour into 100 c.c. of distilled water. Boil the solution in a round-bottomed flask fitted with an air condenser (see p. 46) and use a sand-bath for heating. After boiling for an hour, cool and neutralize the solution with solid calcium carbonate. Add a little water if necessary and filter. Test the filtrate with the following tests:

(a) Make the osazone [see Expt. 41 (d)]. Note that crystals of glucosazone are formed.

(b) Add a little Fehling's solution and boil. Note that reduction takes place.

Instead of using filter-paper, the above experiment may also be carried out with cotton-wool or threads from white cotton material.

Ligno-cellulose. As the cells in plants grow older the walls usually become lignified, that is part of the cellulose becomes converted into ligno-cellulose. The extreme amount of change is found in wood. The least amount in such fibres as those from the stem of the Flax (*Linum usitatissimum*) which, when freed from such impurities, consist of cellulose only and constitute linen. Other fibres, containing more ligno-cellulose, are those of the stem of the Hemp plant (*Cannabis sativa*) and the Jute plant (*Corchorus*) from which string, rope, canvas, sacking and certain carpets are made. The percentages of pure cellulose in these various lignified tissues are as follows:

Cotton fibre	88·3 %
Flax and Hemp fibre ...	72–73 %
Jute	54 %
Beech and Oak wood ...	35–38 %

The ligno-celluloses are generally regarded as consisting of cellulose and two other constituents, of which one contains an aromatic nucleus and the other is of the nature of a pentosan (see xylan, p. 56). Both are sometimes classed together and termed lignin or lignon. The lignin reactions (see below) depend on the presence of an aromatic complex. It has been suggested that coniferin, vanillin and allied compounds which are present in wood are probably the substances responsible for the reaction (Czapek, 8).

Although the best paper is made from cellulose, cheaper forms or paper are manufactured from ligno-cellulose, and, as a result, they give reactions for lignin and are also turned yellow by exposure to light.

Expt. 73. Reactions of lignin.

One of the most striking reactions of lignin (due as it is supposed to a furfural grouping) is the magenta-red coloration given by phloroglucinol in the presence or concentrated hydrochloric acid.

Soak the tissue to be experimented upon with an alcoholic solution of phloroglucinol and then add a drop or two of strong hydrochloric acid. The magenta-red colour will be produced.

As material, practically any lignified tissue may be used. Shavings from twigs of any tree or shrub, e.g. pith and wood from the Elder (*Sambucus nigra*), will be found useful: also shavings from a match; straw, bran, coarse string, cheap white paper, such as newspaper or white and pale-coloured papers used for wrappings.

Make the phloroglucinol test on good white writing paper. It should not give the reaction since it is made from cellulose.

Other phenols (resorcinol, orcinol, catechol, pyrogallol, etc.) and their derivatives will also give colour reactions with lignin in the presence of hydrochloric acid, but the colorations in most cases are not so much developed as with phloroglucinol (Czapek, 8).

It should be noted that strong hydrochloric acid alone will sometimes give a red colour with woody tissues: this is due to the presence or phloroglucinol in the wood itself (see phloroglucinol, p. 102).

Pecto-cellulose. The non-cellulose constituents in this case belong to the class of pectic substances which have already been considered (see p. 65). Such celluloses occur in the cell-walls of the tissues of many fleshy roots, stems and fruits.

Adipo- or cuto-celluloses. These terms have been used for products found in the walls of corky tissue (periderm) and cuticularized tissue (cuticle). More correctly these substances should be termed respectively suberin and cutin, and there is evidence (Priestley, 27) that cellulose is absent from the actual layers of the cell-wall in which suberin and cutin are present. Suberin may be regarded as an aggregate of various condensation products, or anhydrides, of certain organic acids (the suberogenic acids), accompanied by small quantities of glycerides (true fats) of these same acids. By saponification of the condensation products with alkali, three suberogenic acids have been isolated in a more or less pure state, i.e. phellonic acid, $C_{22}H_{42}O_3$, phloionic acid, $C_{22}H_{40}O_7$ and suberinic acid, $C_{17}H_{30}O_3$. The acids themselves are soluble in the usual solvents for fats; phellonic acid or some of its salts may be soluble or have a tendency to swell in water. The anhydrides, on the contrary, are insoluble in solvents for fats and are totally unaffected by water.

There is reason to believe that cutin is an aggregate of similar modifications of various "cutinogenic" acids. The suberin and cutin of one plant probably differ from that of another in the kind and proportion of the acids present.

Hemi-celluloses. These are not strictly celluloses since they are built up of mannans, galactans and pentosans on lines which have already been considered (see pp. 61 and 62). They frequently occur united with each other, for instance as galacto-, gluco- and fructomannan, galactoaraban, galactoxylan, etc. They are found in the cell-walls of the tissues of many seeds, and are apparently hydrolyzed by certain enzymes, termed cytases, during germination.

THE SYNTHESIS AND INTER-RELATIONSHIPS OF CARBOHYDRATES IN THE PLANT.

Now that the properties and characteristics of various carbohydrates have been dealt with, their synthesis and their relationships, one to another, may be considered.

In the previous chapter it has been shown how the plant synthesizes a sugar from carbon dioxide and water by virtue of the chemical energy obtained from transformation of radiant energy by means of chlorophyll. When this sugar reaches a certain concentration in the cell, in the majority of plants, starch is synthesized from it by condensation with elimination of water. The starch is thus the first visible product of assimilation and is temporarily "stored" in an insoluble form during the day, when photosynthesis is active. During the night photosynthesis ceases but the sugar is still translocated from the leaf, as it was in fact during the day; thus, since the supply ceases, the concentration in the cell falls, and the "stored" starch is then hydrolyzed again into sugar, and the process continues until the leaf is either starch-free, or contains considerably less starch. During the next day, the starch formation is repeated and so forth. The process of hydrolysis of starch is carried out by the enzyme, diastase, with the formation of dextrin and maltose. In all probability this same enzyme controls the synthesis of starch.

On the other hand, it has been shown that many plants do not form starch at all in their leaves but only sugar. Examples are the adult Mangold plant (*Beta vulgaris*) and many Monocotyledons (*Allium, Scilla*).

As to the question of which sugars are present in the leaf, there is only evidence from accurate work on a few plants. Careful investigations have been made of the sugars in leaves of the Mangold (*Beta vulgaris*) (Davis, Daish and Sawyer, 17), Garden Nasturtium (*Tropaeolum majus*) (Brown and Morris, 6), the Snowdrop (*Galanthus nivalis*) (Parkin, 26), the Potato (*Solanum tuberosum*) (Davis and Sawyer, 19) and the Vine (*Vitis vinifera*). The general conclusions drawn from these investigations are that sucrose, glucose, and laevulose are always present in leaves: that maltose results from the hydrolysis of starch, being absent from leaves which do not form starch. Maltose is not present in appreciable quantity even in starch-producing leaves because it is rapidly hydrolyzed into glucose by maltase. (In such cases where it has been detected it has been due to diastase action during the drying of leaves before extraction.) Other leaf carbohydrates are the pentoses which have been found in a good many species examined and may be widely distributed; the pentosans, their condensation products, also occur as well as dextrin (Potato).

The next question to be considered is what sugar is first synthesized in the leaf. Is it glucose, laevulose, sucrose or maltose? It is known that the enzymes, invertase and maltase, are commonly present in leaves and that these enzymes respectively control the hydrolysis, of cane-sugar into glucose and laevulose, and of maltose into glucose. It is also possible that they respectively control the synthesis of sucrose and maltose. Laevulose, likewise, as may be supposed, can be obtained from glucose. Thus all the sugars can be readily converted one into another, but to ascertain which is the first product of synthesis is not an easy problem.

In addition to the above-mentioned work on the nature of the sugars present in leaves, a good deal of careful analysis has been made as to the proportions in which the sugars occur relatively to each other during stated periods of time, with a view to answering the question as to which is the first-formed sugar. There are two possibilities: one, that it is sucrose and that it is hydrolyzed into glucose and fructose: the other, that it is glucose, from which fructose is derived, and the two are then synthesized to form sucrose.

Opinion is divided on this point and there is not at present sufficient experimental evidence to decide the matter. The majority of investigators regard sucrose as the first-formed sugar, and suggest that it is inverted into hexoses for purposes of translocation, since the smaller molecules would diffuse faster. There is experimental evidence that there is an increase in hexoses in the conducting tissues. Others favour the view that glucose is the first-formed sugar, and bring forward evidence to this effect. There is however no reason why hexoses should not be formed first and then converted into cane-sugar and temporarily stored as such, being again reinverted into hexoses for translocation. Nor is there any reason for supposing that the first formed sugar is always the same in every plant.

There appears to be very little doubt that maltose is formed in the hydrolysis of starch, and also that starch is a temporary reserve material in the leaves, but whether formed direct from sucrose or from hexoses cannot be stated.

There is some evidence in favour of the view that glucose is more readily used in respiration than laevulose, for under circumstances when neither can be increased, the glucose tends to disappear.

From the leaf the various sugars are translocated to other organs of the plant, e.g. root, stem, flower, fruit and seed. In some cases starch is synthesized from the sugars and "stored" in roots, tubers, tuberous stems, fruits and seeds. In other cases the sugars themselves may be "stored," as, for instance, in the root of the Beet (*Beta vulgaris*), or they may have a biological significance, as in sweet fruits. It must also be borne in mind that sugars are employed throughout the plant in respiration and in the synthesis of more complex substances, i.e. cellulose, gums, pentosans, mucilage, aromatic substances, fats and to a certain extent proteins: in fact they or their precursors constitute the basis from which all organic compounds are synthesized.

The following experiments can be performed with either the Garden Beet or the Mangold Wurzel, both of which are varieties of *Beta vulgaris*, the Common Beetroot. The sugars in the leaves and petioles of the Mangold have been investigated (Davis, Daish and Sawyer, 17) and sucrose, laevulose and glucose have been found. Starch is absent in the adult plant and also maltose. The opinion is held that sucrose is the first-formed sugar of photosynthesis and that this is hydrolyzed for translocation on account of the greater rate of diffusion of the smaller molecules of glucose and laevulose. These are again synthesized in the root to form sucrose where the latter is stored, and hexoses are almost absent from this organ. Though the facts concerning the distribution of the sugars stated above are reliable, it is not certain that the deductions are permissible. The leaf contains the enzymes, invertase, maltase and diastase (Robertson, Irvine and Dobson, 28).

In connexion with the occurrence of various sugars in leaves it is of interest to note that glucose, fructose and mannose can pass over into one another in alkaline aqueous solution. This has been explained by their conversion into the enolic (unsaturated) form common to all three hexoses:

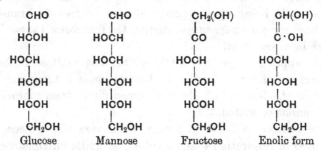

| CHO | CHO | CH₂(OH) | CH(OH) |

Glucose   Mannose   Fructose   Enolic form

Expt. 74. Formation of laevulose and mannose from glucose by alkalies. Into a small flask put 50 c.c. of a 5 % solution of glucose and add 5 c.c. of a 10 % solution of potash. Cork the flask and leave it in an incubator at 35° C. for 24 hours. Cool and neutralize the alkali with a few drops of acetic acid. Test a few drops for laevulose with Seliwanoff's reaction (see p. 53). Then add 5 c.c. of a solution of phenylhydrazine acetate (5 gms. of phenylhydrazine dissolved in 5 c.c. of glacial acetic acid) and shake well. After a few seconds, the solution becomes turbid and a precipitate of mannose-hydrazone is formed (see p. 52). Examine under the microscope and note the characteristic crystalline spheroids.

It is not known whether the pentoses are formed *de novo* in carbon assimilation or whether they arise from the hexoses. A relationship of interest in this connexion is that between the pentoses and the hexoses actually occurring in plants, as will be seen by comparing the formula of *d*-xylose with that of *d*-glucose and the formula of *l*-arabinose with that of *d*-galactose. An additional interest lies in the fact that galactose and arabinose occur so frequently together in gums, while other polysaccharides give glucose and xylose on hydrolysis. (See also Appendix, p. 183.)

Expt. 75. To show the presence of both hexoses and sucrose in the leaf (Davis, Daish and Sawyer, 17). Take about 5 gms. of fresh leaf of either the Beet or Mangold. [Leaves of the Garden Nasturtium (*Tropaeolum majus*) and Wild Chervil (*Chaerophyllum sylvestre*) may also be used.] Tear them into small pieces and drop them into boiling 90–98 % alcohol in a flask on a water-bath. In this way the enzymes of the leaf are killed, and no changes will occur in the carbohydrates present. After boiling for a short time, the alcohol is filtered off and the extraction repeated. Evaporate the filtrate to dryness in an evaporating dish on a water-bath. The filtrate will contain chlorophyll and various pigments, sugars, glucosides, aromatic compounds and other substances according to the plant used. Then add about 20 c.c. of water and at intervals a few drops of basic lead acetate until it ceases to form a precipitate. By this means all hexoses *combined* with aromatic substances as glucosides (see p. 157) are precipitated as insoluble lead salts. The precipitate is filtered off and the lead in the filtrate removed by 1 % sodium carbonate, avoiding excess. Filter again and the filtrate will contain the sugars. Boil the latter and add Fehling's solution drop by drop till reduction ceases. Filter off the copper oxide and then boil the solution with dilute sulphuric acid for a few minutes and make neutral to litmus. Reduction will occur on adding more Fehling and boiling, owing to the inversion of the cane-sugar present.

Expt. 76. *To show the presence of hexoses in the leaf by means of the formation of glucosazone.* Leaves of *Beta, Chaerophyllum sylvestre,* or *Tropaeolum* may be used. Extract as in the previous experiment and precipitate the glucosides with the minimal amount of basic lead acetate. Test for osazone in the filtrate as in Expt. 41 (*d*).

Expt. 77. *To obtain starch from green leaves.* Weigh out 25 gms. of leaflets of the Pea (*Pisum sativum*). The leaves should have been picked in the evening after a sunny day, and it does not matter if the cut leaves are left overnight. Dip the leaflets for a moment into boiling water, remove excess of water and drop them into 200 c.c. of 96–98 % alcohol and boil till the chlorophyll is extracted: then filter. Take the residue of leaves and pound (but not finely) in a mortar and then wash thoroughly with distilled water. Filter through muslin and press free from water (this process extracts most of the protein). Boil the residue with 100 c.c. of water and filter. To the filtrate add iodine. At first the colour may disappear owing to the presence of protein in solution in addition to the starch. When more iodine is added a deep blue coloration is formed.

PLANT ENZYMES WHICH HYDROLYZE CARBOHYDRATES.

Diastase. In the plant starch may be regarded as a reserve product. It is synthesized from sugar, and may be again hydrolyzed into sugar. It can be shown experimentally that starch is converted into glucose by boiling with acids, but in the plant the hydrolysis of starch is catalyzed by the enzyme, diastase. Although the reaction is doubtless of considerable complexity, it may, broadly speaking, be represented as follows:

$$(C_6H_{10}O_5)_n + H_2O \longrightarrow (C_6H_{10}O_5)_x + C_{12}H_{22}O_{11}$$
$$\text{Dextrin} \qquad \text{Maltose}$$

Thus the final products under these conditions are dextrin and the disaccharide, maltose; and not glucose.

It is reasonable to assume that cells which contain starch also either contain, or are capable of producing, diastase. But the amount of diastase present, or at any rate capable of being extracted, varies in different tissues. Diastase, like most enzymes, is soluble in water. In many cases, however, a water-extract from fresh crushed tissues in which diastase occurs, will not contain any appreciable amount of enzyme. This is sometimes due to the fact that the protoplasm does not readily yield up the enzyme until it has been killed. If the tissues are dried at a moderate temperature (30–40° C.) both the powdered leaves themselves and a water extract are fairly rich in diastase; or, if the living tissues are macerated and extracted with water to which chloroform has been added, the cells die more rapidly and yield up the enzyme to the solvent. From such a water extract, a crude precipitate containing the enzyme may be obtained by addition of alcohol. For obtaining the maximum results with diastatic activity in leaves, a water extract should be made after they have been killed, either by drying, or by the action of toluol or chloroform.

It has been shown (Brown and Morris, 6) that in leaves which contain tannin, the presence of the latter largely inhibits the action of the enzyme and may be the cause, in such cases, of an entire lack of activity in the extract.

The diastatic activity of leaves appears to vary largely in different genera and species. The subject has been investigated (Brown and Morris, 6) and a list of their relative activities has been drawn up as follows.

[The numbers represent the amount of maltose, expressed in grams, which 10 gms. of air-dried leaf will produce from soluble-starch (starch treated with dilute hydrochloric acid) by hydrolysis in 48 hrs. at 30° C.]

Pisum sativum...................240·30	Helianthus annuus............... 3·94
Phaseolus multiflorus.........110·49	H. tuberosus .... 3·78
Lathyrus odoratus ....100·37	Funkia sinensis .... 5·91
L. pratensis .... 34·79	Allium Cepa .... 3·76
Trifolium pratense .... 89·66	Hemerocallis fulva............... 2·07
T. ochroleucum .... 56·21	Populus sp. .... 3·79
Vicia sativa .... 79·55	Syringa vulgaris.................. 2·53
V. hirsuta........................ 53·23	Cotyledon Umbilicus............ 4·61
Lotus corniculatus .... 19·48	Humulus Lupulus ....2·01–9·60
Lupinus sp. .... 3·51	Hymenophyllum demissum ... 4·20
Grass with Clover .... 27·92	Hydrocharis Morsus-ranae ... 0·267
Tropaeolum majus.........3·68–9·64	

From the above table it is seen that the leaves of genera of the Leguminosae are apparently very rich in diastase. Whether this is so, or whether in other plants the diastatic activity is inhibited by other substances, has not yet been ascertained. As mentioned above, tannins inhibit the action of diastase, and hence leaves rich in tannin, e.g. Hop (*Humulus*), cannot be expected to yield good results.

The tissues of germinating barley (*Hordeum vulgare*) also contain large quantities of diastase, and this material can be used to demonstrate the solubility, isolation and activity of the enzyme.

The action on starch of diastase from the leaf of the Common Pea (*Pisum sativum*) and from germinating barley grains has already been demonstrated [see Expt. 57 (a) and (b)] in connexion with dextrin. The following experiments have special reference to the enzyme.

Expt. 78. To demonstrate the activity of diastase from germinating barley. (See also Expt. 57.)

Pound up 2–3 gms. of germinated barley grains in a mortar and extract the mass with 50 c.c. of water. Filter, and take two equal portions in two test-tubes. Boil one tube. To both tubes add an equal quantity of the starch solution prepared as in Expt. 55. Place the tubes in a beaker of water at 38–40° C. From time to time withdraw a drop from each tube with a pipette and test with iodine solution on a white tile. The starch in the unboiled tube will gradually give the dextrin reactions (see p. 59); that in the boiled tube will remain unchanged.

This simple method may also be adopted for showing the diastatic activity of leaves. Instead of germinating barley, a few leaflets of the Pea (*Pisum sativum*) or Clover (*Trifolium pratense*) should be pounded up in a mortar and extracted with 50 c.c. of water and filtered.

Maltase. This enzyme hydrolyzes maltose into two molecules of glucose:

$$C_{12}H_{22}O_{11} + H_2O = 2C_6H_{12}O_6.$$

Investigations upon maltase have, until recently, produced rather contradictory results, but later work (Davis, 14: Daish, 15, 16) has led to more satisfactory conclusions. The latter show that maltase is most probably present in all plants in which hydrolysis of starch occurs. It has been detected in leaves of the Nasturtium (*Tropaeolum*), the Potato (*Solanum*), the Dahlia, the Turnip (*Brassica*), the Sunflower (*Helianthus*) and the Mangold (*Beta*), and it is most probably widely distributed in foliage leaves. Its detection is not easy for various reasons which are as follows. It is not readily extracted from the tissues by water: it is unstable, being easily destroyed by alcohol and chloroform. Its activity is also limited or even destroyed at temperatures above 50° C. Hence the extraction of maltase, by merely pounding up tissues with water, does not yield good results: moreover, as an antiseptic, toluol must be used and not chloroform. Finally, if the enzyme is to be extracted from dried material, this must not be heated at too high a temperature previous to the extraction.

Maltase occurs in quantity in both germinated and ungerminated seeds of cereals. If, in kilning, malt has not been heated at too high a temperature, the maltase may not be destroyed, and, in such cases, malt extract will contain both diastase and maltase. This would explain the fact that glucose, instead of maltose, has sometimes been obtained by the action of malt diastase on starch. In other cases, when a higher temperature has been employed, the maltase will be destroyed. Maltase itself, of course, does not act directly upon starch but only on maltose. The use of chloroform, as an antiseptic, by some observers explains how they came to overlook the presence of maltase, thus obtaining maltose, and not glucose, as an end product in hydrolysis by malt extracts. The optimum temperature for the maltase reaction is 39° C.

The presence of maltase in leaves is not readily shown for the following reasons. Since maltase is destroyed by alcohol, the preparation of a crude precipitate of the enzyme by precipitating a water extract of the leaves is not satisfactory. If the water extract is added directly to maltose, and incubated, hydrolysis may be demonstrated by determining the reducing power of the sugars formed. A control experiment must, however, be made by incubating the water extract alone, and subsequently determining the reducing power of any sugars present.

Invertase. This enzyme hydrolyzes cane-sugar into one molecule of glucose and one molecule of laevulose:

$$C_{12}H_{22}O_{11} + H_2O = C_6H_{12}O_6 + C_6H_{12}O_6.$$

Invertase is probably very widely distributed in plants. Its presence has been demonstrated in the leaves and stem, though not in the root, of the Beet (*Beta vulgaris*) (Robertson, Irvine and Dobson, 28). Also in the leaves of a number of other plants (Kastle and Clark, 22). Its detection, by its action on sucrose, is not easy on account of the presence of other enzymes and reducing sugars in leaf extracts.

The absence of invertase from the root of the Beet raises a difficulty as to how the cane-sugar is synthesized from the hexoses supplied from the leaves (see p. 73). Some observers (Robertson, Irvine and Dobson, 28) incline to the view that cane-sugar is synthesized in the stems and travels as such to the roots. Others (Davis, Daish and Sawyer, 17) maintain that the cane-sugar is synthesized in the root, even though invertase is absent.

REFERENCES

BOOKS

1. **Abderhalden, E.** Biochemisches Handlexikon, II. Berlin, 1911.
2. **Armstrong, E. F.** The Simple Carbohydrates and the Glucosides. London, 1924. 4th ed.
3. **Atkins, W. R. G.** Some Recent Researches in Plant Physiology. London, 1916.
4. **Mackenzie, J. E.** The Sugars and their Simple Derivatives. London, 1913.

PAPERS

5. **Blackman, F. F.** The Biochemistry of Carbohydrate Production in the Higher Plants from the Point of View of Systematic Relationship. *N. Phytol.*, 1921, Vol. 20, pp. 2–9.
6. **Brown, H. T.**, and **Morris, G. H.** A Contribution to the Chemistry and Physiology of Foliage Leaves. *J. Chem. Soc.*, 1893, Vol. 63, pp. 604–677.
7. **Clayson, D. H. F., Norris, F. W.**, and **Schryver, S. B.** The Pectic Substances of Plants. Part II. A Preliminary Investigation of the Chemistry of the Cell-Walls of Plants. *Biochem. J.*, 1921, Vol. 15, pp. 643–653.
8. **Czapek, F.** Ueber die sogenannten Ligninreactionen des Holzes. *Zs. physiol. Chem.*, 1899, Vol. 27, pp. 141–166.
9. **Davis, W. A.**, and **Daish, A. J.** A Study of the Methods of Estimation of Carbohydrates, especially in Plant-extracts. A new Method for the Estimation of Maltose in Presence of other Sugars. *J. Agric. Sci.*, 1913, Vol. 5, pp. 437–468.
10. **Davis, W. A.**, and **Daish, A. J.** Methods of estimating Carbohydrates. II. The Estimation of Starch in Plant Material. The Use of Taka-Diastase. *J. Agric. Sci.*, 1914, Vol. 6, pp. 152–168.
11. **Daish, A. J.** Methods of Estimation of Carbohydrates. III. The Cupric Reducing Power of the Pentoses—Xylose and Arabinose. *J. Agric. Sci.*, 1914, Vol. 6, pp. 255–262.
12. **Davis, W. A.**, and **Sawyer, G. C.** The Estimation of Carbohydrates. IV. The Presence of Free Pentoses in Plant Extracts and the Influence of other Sugars on their Estimation. *J. Agric. Sci.*, 1914, Vol. 6, pp. 406–412.
13. **Davis, W. A.** The Hydrolysis of Maltose by Hydrochloric Acid under the Herzfeld Conditions of Inversion. A Reply to A. J. Kluyver. *J. Agric. Sci.*, 1914, Vol. 6, pp. 413–416.
14. **Davis, W. A.** The Distribution of Maltase in Plants. I. The Function of Maltase in Starch Degradation and its Influence on the Amyloclastic Activity of Plant Materials. *Biochem. J.*, 1916, Vol. 10, pp. 31–48.
15. **Daish, A. J.** The Distribution of Maltase in Plants. II. The Presence of Maltase in Foliage Leaves. *Biochem. J.*, 1916, Vol. 10, pp. 49–55.
16. **Daish, A. J.** The Distribution of Maltase in Plants. III. The Presence of Maltase in Germinated Barley. *Biochem. J.*, 1916, Vol, 10, pp. 56–76.
17 **Davis, W. A., Daish, A. J.**, and **Sawyer, G. C.** Studies of the Formation and Translocation of Carbohydrates in Plants. I. The Carbohydrates of the Mangold Leaf. *J. Agric. Sci.*, 1916, Vol. 7, pp. 255–326.

80   CARBOHYDRATES

18. **Davis, W. A.** Studies of the Formation, etc. II. The Dextrose-Laevulose Ratio in the Mangold. *J. Agric. Sci.*, 1916, Vol. 7, pp. 327–351.

19. **Davis, W. A.,** and **Sawyer, G. C.** Studies of the Formation, etc. III. The Carbohydrates of the Leaf and Leaf Stalks of the Potato. The Mechanism of the Degradation of Starch in the Leaf. *J. Agric. Sci.*, 1916, Vol. 7, pp. 352–384.

20. **Davis, W. A.** The Estimation of Carbohydrates. V. The supposed Precipitation of Reducing Sugars by Basic Lead Acetate. *J. Agric. Sci.*, 1916, Vol. 8, pp. 7–15.

21. **Haynes, D.** The Gelatinisation of Pectin in Solutions of the Alkalies and the Alkaline Earths. *Biochem. J.*, 1914, Vol. 8, pp. 553–583.

22. **Kastle, J. H.,** and **Clark, M. E.** On the Occurrence of Invertase in Plants. *Amer. Chem. J.*, 1903, Vol. 30, pp. 421–427.

23. **Neville, A.** Linseed Mucilage. *J. Agric. Sci.*, 1913, Vol. 5, pp. 113–128.

24. **Parkin, J.** Contributions to our Knowledge of the Formation, Storage and Depletion of Carbohydrates in Monocotyledons. *Phil. Trans. R. Soc.*, B Vol. 191, 1899, pp. 35–79.

25. **Parkin, J.** On a Reserve Carbohydrate which produces Mannose, from the Bulb of *Lilium*. *Proc. Camb. Phil. Soc.*, 1900–1902, Vol. 11, pp. 139–142.

26. **Parkin, J.** The Carbohydrates of the Foliage Leaf of the Snowdrop (*Galanthus nivalis*), and their Bearing on the First Sugar of Photosynthesis. *Biochem. J.*, 1911, Vol. 6, pp. 1–47.

27. **Priestley, J. H.** Suberin and Cutin. *N. Phytol.*, 1921, vol. 20, pp.17–29.

28. **Robertson, R. A., Irvine, J. C.,** and **Dobson, M. E.** A Polarimetric Study of the Sucroclastic Enzymes in *Beta vulgaris*. *Biochem. J.*, 1909, Vol. 4, pp. 258–273.

29. **Salkowski, E.** Ueber die Darstellung des Xylans. *Zs. physiol. Chem.* 1901–2, Vol. 34, pp. 162–180.

30. **Salkowski, E.** Ueber das Verhalten des Arabans zu Fehling'scher Lösung. *Zs. physiol. Chem.*, 1902, Vol. 35, pp. 240–245.

31. **Schryver, S. B.,** and **Haynes, D.** The Pectic Substances of Plants. *Biochem. J.*, 1916, Vol. 10, pp. 539–547.

32. **Schulze, E., Steiger, E.,** und **Maxwell, W.** Zur Chemie der Pflanzenzellmembranen. I. Abhandlung. *Zs. physiol. Chem.*, 1890, Vol. 14, pp. 227–273.

33. **Spoehr, H. A.** The Carbohydrate Economy of Cacti. *Carnegie Institution of Washington Publication*, 1919, No. 287.

34. **Tutin, F.** The Behaviour of Pectin towards Alkalis and Pectase. *Biochem. J.*, 1921, Vol. 15, pp. 494–497.

CHAPTER VI

THE VEGETABLE ACIDS

THOUGH the name "vegetable acids" might strictly be applied to all acids found in plants, it is, as a rule, restricted to certain acids and hydroxy-acids of the methane, ethylene and acetylene series.

We may take first the acids of the methane series which biologically fall into two groups, the simpler members associated with fundamental metabolism and the more complex ones associated with fat formation. The first six members, at least, may be included among the vegetable acids in the narrow sense. They are liquids, readily volatile in steam, and several of them, without doubt, are closely involved in some of the most fundamental and important reactions of plant metabolism. In fact their relationships to certain of the amino-acids which are constituents of most proteins, cannot be too strongly emphasized. The higher members (with ten and more carbon atoms) are solids insoluble in water. The glycerol esters of certain of these higher members are important constituents of the plant fats and will be considered in the following chapter. The first six representatives of the series are:

Acids of the methane series		Corresponding amino-acids
Formic acid	$H \cdot COOH$	
Acetic acid	$CH_3 \cdot COOH$	amino-acetic acid or glycine
Propionic acid	$CH_3 \cdot CH_2 \cdot COOH$	amino-propionic acid or alanine
Butyric acid	$CH_3 \cdot CH_2 \cdot CH_2 \cdot COOH$	
Valeric acid	$CH_3 \cdot CH_2 \cdot CH_2 \cdot CH_2 \cdot COOH$	amino-iso-valeric acid or valine
Caproic acid	$CH_3 \cdot CH_2 \cdot CH_2 \cdot CH_2 \cdot CH_2 \cdot COOH$	amino-iso-caproic acid or leucine

Formic acid can be obtained by submitting plants to steam distillation. This indicates that it probably exists in the free state in plants, though there is the possibility of its being formed from other substances during distillation. There is good evidence (Dobbin, 1), however, that it is present in the stinging hairs of the Nettle (*Urtica dioica*). It is a liquid which is volatile with steam and can be readily reduced to formaldehyde with nascent hydrogen.

Expt. 79. *Tests for formic acid.* Make a solution of formic acid (1 c.c. acid : 100 c.c. water) and perform the following tests :

(*a*) Acidify 10 c.c. with a few drops of strong hydrochloric acid and add a little magnesium powder. The formic acid will be reduced to formaldehyde. Filter and test for the latter by Schryver's test (see p. 39).

(*b*) Neutralize a few c.c. of the solution with dilute caustic soda and add a few drops of 5 % mercuric chloride solution and heat. The mercuric salt is reduced to mercurous chloride which is precipitated, being insoluble.

Expt. 80. *Detection of formic acid in the Nettle* (Urtica dioica). Take a strong filter-paper (about 10 cms. in diameter) of the best quality and soak it in a concentrated solution of barium hydroxide. Allow the paper to dry in air, whereby the barium hydroxide is converted into carbonate. Take at least 200 Nettle leaves, and, with gloved hands, carefully blot both sides of the leaves between the folded paper. Break up the paper in about 40 c.c. of distilled water, warm and filter on the pump. Wash with 10 c.c. of hot water. To the filtrate containing barium formate add 0·5 gm. of glacial phosphoric acid and distil with a water condenser. Add about 20 drops of strong hydrochloric acid to the distillate and then magnesium powder. When hydrogen is no longer evolved, filter and test for formaldehyde by Schryver's reaction. A positive result will be obtained.

Acetic acid has been found to occur in plants, both in the free state and as salts and esters. Possibly, however, in some cases it may have arisen from the decomposition of other substances during distillation.

Propionic acid has rarely been detected in plants. **Butyric, isobutyric** and **caproic acids** have been detected in a few plants.

Isovaleric acid has been isolated from various plants, notably species of Valerian (*Valeriana*).

Esters of the above acids form important plant constituents since they are responsible for many fruit odours. Amyl acetate, for instance, occurs in the fruit of the Banana (*Musa sapientum*): amyl formate, acetate and caproate are probably present in the fruit of the Apple (*Pyrus Malus*), etc. Such compounds are frequently classed with the "essential oils" (see p. 108).

The next group to be considered are the monohydroxy-acids of the methane series. Of these glycollic acid may be mentioned.

Glycollic acid, or hydroxy-acetic acid, $CH_2 \cdot OH \cdot COOH$, has been isolated from unripe fruit of the Grape and from the leaves of the Virginian Creeper (*Ampelopsis hederacea*). Also from the Sugar-cane (*Saccharum officinarum*), the Lucerne (*Medicago sativa*) and the Tomato (*Lycopersicum esculentum*). Its relationship to the amino-acid, glycine (see p. 134), should be borne in mind.

The dibasic acids of the methane series contain several important members:

Dibasic acids	Corresponding amino-acids
Oxalic acid $(COOH)_2$	
Malonic acid $CH_2 \cdot (COOH)_2$	
Succinic acid $CH_2 \cdot CH_2 \cdot (COOH)_2$	amino-succinic or aspartic acid
Glutaric acid $CH_2 \cdot CH_2 \cdot CH_2 \cdot (COOH)_2$	amino-glutaric or glutaminic acid
Adipic acid $CH_2 \cdot CH_2 \cdot CH_2 \cdot CH_2 \cdot (COOH)_2$	

Oxalic acid occurs very frequently and widely distributed in plants, usually as the calcium salt, and apparently less frequently as the sodium and potassium salts. It has rarely been detected as the free acid. It is especially abundant in spp. of *Oxalis*, in the Rhubarb (*Rheum Rhaponticum*) and Sorrel (*Rumex Acetosa*). The calcium salt is precipitated on adding calcium acetate to a solution of the acid. Calcium oxalate is insoluble in acetic acid, but soluble in dilute mineral acids.

Expt. 81. *Tests for oxalic acid.* Take a 2% solution of oxalic acid, neutralize with caustic soda (or use a soluble oxalate) and make the following tests:

(*a*) To 5 c.c. add a few drops of 5% calcium chloride solution. A white precipitate of calcium oxalate is formed. Divide the precipitate into two portions. To one add an equal quantity of strong acetic acid: the precipitate is insoluble even on heating. To the other add strong hydrochloric acid drop by drop: the precipitate is soluble. Hence the free acid can be precipitated with calcium acetate but not with calcium chloride.

(*b*) To 5 c.c. add a few drops of 5% lead acetate solution. A white precipitate of lead oxalate is formed. Add an equal quantity of strong acetic acid and warm; the precipitate is insoluble.

Expt. 82. *Preparation of calcium oxalate from leaves of the Sorrel* (Rumex Acetosa). Take 100 gms. of fresh leaves of the Sorrel. Boil them in an evaporating dish with 200 c.c. of water and squeeze the boiled mass through linen. Boil the filtrate again and filter on a pump. Acidify the filtrate with acetic acid, and add a concentrated solution of calcium acetate until no more precipitate is formed. The precipitate cannot readily be filtered off so that it should be allowed to settle for 12 hours. Then decant off the liquid and boil up the precipitate in the minimum amount of 10% hydrochloric acid. On cooling, calcium oxalate will separate out in characteristic crystals. On examining under the microscope, these will be seen to be octahedra, giving the appearance of a square with a diagonal cross (envelope form). Leaves of Rhubarb (*Rheum Rhaponticum*) can also be used, taking about 250 gms. in 500 c.c. of water.

It is stated that there is an enzyme widely distributed in plants (Staehelin, 3) which has the power of decomposing oxalic acid with the production of carbon dioxide.

Malonic acid has been isolated from the Sugar Beet (*Beta vulgaris* var. *Rapa*)[1]. It forms insoluble calcium and lead salts.

Succinic acid is probably widely distributed in plants. It has been isolated from the unripe Grape, from fruit of the Gooseberry, Currant, Apple and Banana, from Rhubarb (*Rheum Rhaponticum*), Greater Celandine (*Chelidonium majus*) and other plants. Succinic acid crystallizes well in rhombic prisms or plates. It is not very readily soluble in cold water, though more so in hot. Its salts with the alkali metals are readily soluble. Calcium succinate is deposited as a crystalline precipitate on adding calcium chloride to fairly concentrated solutions of the acid after neutralization (or of a soluble succinate), but from a dilute solution it is not precipitated except on addition of alcohol. Barium succinate comes down as a crystalline precipitate even from dilute solutions. Ferric succinate is insoluble and its formation is used in the detection of the acid.

The relationship of succinic acid to aspartic, or *a*-amino-succinic, acid which is an abundant constituent of many proteins (see p. 134) should be noted.

Expt. 83. *Tests for succinic acid.* *A.* Take a 1% solution of succinic acid, neutralize with caustic soda (or use a soluble succinate) and make the following tests:

(*a*) To 5 c.c. add a few drops of 5% calcium chloride solution. A slight precipitate is formed, especially on rubbing the sides of the tube with a rod. To another 5 c.c. add again calcium chloride solution followed by an equal volume of 96% alcohol. A white precipitate of calcium succinate is formed.

(*b*) To 5 c.c. add a few drops of 5% barium chloride solution. A crystalline precipitate of barium succinate is formed and, again, its appearance is hastened by rubbing the sides of the tube.

(*c*) To 5 c.c. add a few drops of 5% lead acetate solution. A white precipitate of lead succinate is formed. Add an equal quantity of strong acetic acid. The precipitate is soluble.

(*d*) To 10 c.c. add about 1-2 c.c. of 5% ferric chloride solution. A red-brown gelatinous precipitate of ferric succinate is formed. Filter off the precipitate, wash well and boil with about 20 c.c. of dilute ammonia. Filter off the ferric hydroxide, and to the filtrate, after boiling off any excess of ammonia, add 5% barium chloride solution. A crystalline precipitate of barium succinate is formed. This test constitutes a method for identifying succinic acid.

B. Make a cold concentrated solution of succinic acid, neutralize (or use a soluble succinate) and add 5% calcium chloride solution. A crystalline precipitate of calcium succinate will separate out. Its appearance may be hastened by rubbing or shaking.

[1] It should be noted that an exceptionally large number of chemical substances have been isolated from the Sugar Beet on account of their accumulation in the waste products from sugar manufacture. There is little doubt that the same substances could be isolated from other plants if sufficient quantity of material were employed.

Glutaric and **adipic** acids have been detected in extracts from the root of the Sugar Beet (*Beta vulgaris* var. *Rapa*). It is probable that they also occur in other plants. The relationship of glutaric acid to glutaminic acid is important (see p. 134).

Of the monohydroxy-dibasic acids, malic acid is the best known.

Malic acid. It should be noted that in constitution malic acid is a hydroxy-succinic acid. It is widely distributed in plants, being found in many fruits, such as those of the Apple, Pear, Cherry, etc.; also in leaves and vegetative parts, especially in some succulents (Crassulaceae, *Mesembryanthemum*).

Malic acid crystallizes in colourless needles which are very deliquescent and hence difficult to obtain. Its salts with the alkali metals are soluble. Calcium malate is only precipitated from a very concentrated solution of the acid (after neutralization) or of a soluble malate. Very few well-defined tests can be made for malic acid.

Expt. 84. *Tests for malic acid. A.* Take a 2 % solution of malic acid, neutralize with caustic soda (or use a soluble malate) and make the following tests:

(*a*) Add a few drops of 5 % calcium chloride solution. No precipitate is formed, but the addition of an equal volume of 96 % alcohol will bring down a precipitate of calcium malate.

(*b*) Add a few drops of 5 % lead acetate solution. A white precipitate of lead malate is formed. Add a little acetic acid and warm. The precipitate dissolves.

B. Heat a little solid malic acid in a dry test-tube. It melts and then gives off fumes of maleïc acid which condense in white crystals on the cooler parts of the tube.

Expt. 85. *Preparation of malic acid from apples.* Take six apples (total weight from 500–700 gms.). Cut them into thin slices and drop the slices as quickly as possible into the minimum amount of boiling alcohol in a conical flask. In this way the oxidizing enzymes are destroyed, and brown oxidation products are avoided. After well boiling, filter through paper. Neutralize the filtrate to litmus with sodium hydroxide solution, and add concentrated calcium chloride solution until a precipitate ceases to be formed. Allow the precipitate of calcium malate to settle and then add alternately a few drops of calcium chloride solution and a little alcohol to ensure complete precipitation. Decant, and filter off the calcium malate. Dissolve the malate in the minimum amount of hot water, filter and add concentrated lead acetate solution until a precipitate of lead malate ceases to be formed. Filter off the lead malate, suspend it in a minimum amount of water, and pass in sulphuretted hydrogen until the malate is decomposed. Filter and concentrate in a crystallizing dish on a water-bath. Crystals of malic acid are deposited. Test as in Expt. 84.

Of the dihydroxy-dibasic acids, tartaric acid is the best known. It should be noted that tartaric acid is dihydroxy-succinic acid. Thus the three acids are related as follows:

$$\text{Succinic acid } COOH \cdot CH_2 \cdot CH_2 \cdot COOH$$
$$\text{Malic acid } \quad COOH \cdot CHOH \cdot CH_2 \cdot COOH$$
$$\text{Tartaric acid } COOH \cdot CHOH \cdot CHOH \cdot COOH$$

Tartaric acid is widely distributed in plants, often in the form of the calcium or potassium salts. It occurs in many fruits, as for instance, those of the Grape (*Vitis vinifera*), Tomato (*Lycopersicum esculentum*), Mountain Ash (*Pyrus Aucuparia*) and Pineapple (*Ananas sativus*); it has also been detected in the leaves and vegetative parts of many plants. Tartaric acid is easily soluble in water from which it crystallizes in colourless prisms. Calcium tartrate is only slightly soluble in cold water, though more so in hot. On adding calcium chloride to a soluble tartrate, calcium tartrate is precipitated, more or less rapidly according to the strength of the solution, and sometimes as a crystalline precipitate. The crystals may occur as characteristic rhombic prisms with octahedral faces or as needles. The precipitate is soluble in acetic acid. The acid potassium salt of tartaric acid is soluble with difficulty in water and hence is used in identification of the acid.

Racemic acid, which is a combination of dextro- and laevo-tartaric acids, is also found in certain varieties of the Grape. Calcium racemate is insoluble in acetic acid. It is soluble in hydrochloric acid from which it separates out rapidly in a crystalline state on neutralizing with ammonia (tartrate only separates slowly).

Expt. 86. *Tests for tartaric acid. A.* Take a 1% solution of tartaric acid, neutralize with caustic soda (or use a soluble tartrate) and make the following tests:

(*a*) Add a few drops of 5% calcium chloride solution. A white precipitate of calcium tartrate is formed. Add an equal volume of glacial acetic acid and warm; the precipitate dissolves.

(*b*) Add a few drops of 5% lead acetate solution. A white precipitate of lead tartrate is formed. Add acetic acid and warm; the precipitate dissolves.

(*c*) To 2–3 c.c. in a test-tube add a few drops of ferrous sulphate solution. Place the test-tube in a beaker of cold water, and add a few drops of hydrogen peroxide followed by an excess of caustic soda solution. A deep violet or blue colour is obtained. The colour is due to the formation of dihydroxymaleïc acid and the reaction of this with the ferric salt present.

(*d*) To one drop of tartrate add 2 drops of a 2% solution of resorcinol and then 3 c.c. of strong sulphuric acid. Heat gently. A rose colour is formed which deepens to a violet-red.

B. Take 2–3 cm. of a strong solution of tartaric acid, acidify with glacial acetic acid and add a little potassium acetate solution. A white crystalline precipitate of potassium hydrogen tartrate will be formed.

Expt. 87. Identification of tartaric acid in grapes. Take 150–200 gms. of unripe grapes (early July) and boil them well with the minimum amount of water in an evaporating dish. As they soften they should be well stirred and crushed. Then filter and squeeze the mass through strong linen. Neutralize the filtrate with caustic soda, heat to boiling and filter on a pump. Cool the filtrate, and add 2–3 c.c. of saturated calcium chloride solution. Allow the mixture to stand for 24 hours. A crystalline precipitate will separate out. Under the microscope this will be seen to consist of needles and octahedra. The needles are a double salt of *d*-tartaric and *l*-malic acid (Ordonneau, 2); the octahedra consist either of tartaric acid or racemic acid or a mixture of both. Filter off this precipitate and heat in 50 % acetic acid. The double salt and the tartaric acid will dissolve, but octahedra of racemic acid (if present) will remain undissolved. Filter and make the following tests with the filtrate:

(*a*) Add to a small quantity in a test-tube, resorcinol and sulphuric acid as in Expt. 86 *A* (*d*); a positive result is given.

(*b*) Evaporate down the remainder on a water-bath and add potassium acetate and acetic acid as in Expt. 86 *B*; potassium hydrogen tartrate crystallizes out.

If octahedra are left undissolved after treating with 50 % acetic acid, racemic acid is present. Heat this residue with dilute hydrochloric acid. It will go into solution. Neutralize a portion with ammonia, and the acid will crystallize out at once. Test another portion with resorcinol as in (*a*); a positive result will be given.

Of the tribasic acids, citric acid, $C_3H_4 \cdot OH \cdot (COOH)_3$, is the most important.

Citric acid occurs in large quantities in fruits of the genus *Citrus*, i.e. in the Orange, Lime, Lemon, etc. Also in many other fruits, such as the Gooseberry, Currant, Tomato, etc.

Expt. 88. Tests for citric acid. A. Take a 1 % solution of citric acid, neutralize it with caustic soda (or use a soluble citrate) and make the following tests:

(*a*) Add 5 % calcium chloride solution. No precipitate is given. Heat to boiling and a white precipitate of calcium citrate is formed. Calcium citrate is soluble in cold water but insoluble in hot water.

(*b*) Add 5 % lead acetate solution. A white precipitate of lead citrate is formed. Add an equal quantity of acetic acid and warm: the precipitate is soluble.

B. Take 5 c.c. of a 2 % solution of citric acid and add 3 c.c. of Denigès' reagent (prepared by dissolving with the aid of heat 1 gm. of mercuric oxide in a mixture of 4 c.c. of strong sulphuric acid and 20 c.c. of distilled water). Boil, and add a 2 % solution of potassium permanganate drop by drop. The permanganate is at first decolorized, but, on further cautious addition, the colour persists. Finally the liquid becomes turbid and a white precipitate forms. This is due to a mercury compound of acetone-dicarboxylic acid, resulting from the oxidation of citric acid by the permanganate.

C. Heat gently a few crystals of citric acid for some time with an equal weight of resorcinol and a few drops of concentrated sulphuric acid. Add excess of alkali; the solution shows a fine blue fluorescence due to the presence of a product, resocyan.

Expt. 89. *Preparation of citric acid from lemons.* Squeeze the juice from three lemons and filter through muslin. Measure the volume of the juice, and add strong caustic soda solution, carefully, until the reaction is slightly alkaline. Filter and for every 10 c.c. of juice, add 5 c.c. of a 10 % solution of calcium chloride. No precipitate is formed. Now heat to boiling and a copious precipitate of calcium citrate is formed. Filter off, while hot, on a filter-pump, wash with a little boiling water, drain well and dry in the air. Weigh and add the requisite amount of sulphuric acid (1 gm. of citrate = 15 c.c. of normal sulphuric acid). Allow the mixture to stand for a short time, filter and concentrate the filtrate in a glass dish on a water bath. Crystals of citric acid separate out on concentrating considerably. (If, for any reason, insufficient sulphuric acid has been added, some calcium citrate may separate out first on concentrating. If so, add a few drops of sulphuric acid, filter and continue to concentrate.) Drain off the citric acid on a filter-pump, dissolve in water and make the tests in Expt. 88.

The acids of the ethylene series have not as yet been very widely detected.

Fumaric acid, $COOH \cdot CH = CH \cdot COOH$, occurs in the Fumariaceae (*Fumaria, Corydalis*) and Papaveraceae (*Glaucium*).

Aconitic acid, $COOH \cdot CH_2 \cdot C \cdot COOH \cdot CH \cdot COOH$, is found in the Monkshood (*Aconitum*) and other genera of the Ranunculaceae.

The best known acid of the acetylene series is **sorbic acid**, found in berries of the Mountain Ash (*Pyrus Aucuparia*).

REFERENCES

1. **Dobbin, L.** On the Presence of Formic Acid in the Stinging Hairs of the Nettle. *Proc. Roy. Soc.*, Edinburgh, 1920, Vol. 39, pp. 137–142.
2. **Ordonneau, Ch.** De l'acidité des raisins verts et de la préparation de l'acide malique. *Bull. de la soc. chim.*, 1891, Vol. 6, pp. 261–264.
3. **Staehelin, M.** Die Rolle der Oxalsäure in der Pflanze. Enzymatischer Abbau des Oxalations. *Biochem. Zeitschr.*, 1919, Vol. 96, pp. 1–49.

CHAPTER VII

FATS AND ALLIED SUBSTANCES

A FAT may be defined as an ester or glyceride of a fatty acid. Just as an inorganic salt, such as sodium chloride, is formed by the reaction of hydrochloric acid with sodium hydroxide, so a fat is formed by the reaction of the trihydric alcohol, glycerol, and a fatty acid.

The word *fat* is not a familiar one in botanical literature, the term *oil* being more commonly used. It is generally met with in connexion with the reserve products of seeds. The oils of seeds are, however, true fats. The term oil may be misleading to some extent, because a fat which is liquid at ordinary temperatures is usually spoken of as an oil, and yet there are also many other substances, of widely differing chemical composition, which have the physical properties of oils, and which are known as such.

Most of the vegetable fats are liquid at ordinary temperatures but some are solids.

The best-known series of acids from which fats are formed is the series $C_nH_{2n}O_2$ of which formic acid is the first member. The other members of the series are:

Formic acid	$H \cdot COOH$	[1] Myristic acid	$C_{13}H_{27} \cdot COOH$
[1] Acetic acid	$CH_3 \cdot COOH$	Isocetic acid	$C_{14}H_{29} \cdot COOH$
Propionic acid	$C_2H_5 \cdot COOH$	[1] Palmitic acid	$C_{15}H_{31} \cdot COOH$
[1] Butyric acid	$C_3H_7 \cdot COOH$	Daturic acid	$C_{16}H_{33} \cdot COOH$
Valeric acid	$C_4H_9 \cdot COOH$	[1] Stearic acid	$C_{17}H_{35} \cdot COOH$
[1] Caproic acid	$C_5H_{11} \cdot COOH$	Nonadecylic acid	$C_{18}H_{37} \cdot COOH$
Œnanthylic acid	$C_6H_{13} \cdot COOH$	[1] Arachidic acid	$C_{19}H_{39} \cdot COOH$
[1] Caprylic acid	$C_7H_{15} \cdot COOH$	[1] Behenic acid	$C_{21}H_{43} \cdot COOH$
Pelargonic acid	$C_8H_{17} \cdot COOH$	[2] Lignoceric acid ⎫	
[1] Capric acid	$C_9H_{19} \cdot COOH$	[2] Carnaübic acid ⎬	$C_{23}H_{47} \cdot COOH$
Undecylic acid	$C_{10}H_{21} \cdot COOH$	Hyaenic acid	$C_{24}H_{49} \cdot COOH$
[1] Lauric acid	$C_{11}H_{23} \cdot COOH$	[2] Cerotic acid	$C_{25}H_{51} \cdot COOH$
Tridecylic acid	$C_{12}H_{25} \cdot COOH$	[2] Melissic acid	$C_{29}H_{59} \cdot COOH$

[1] Occur in fats. [2] Occur in waxes.

Another series is the oleic or acrylic series $C_nH_{2n-2}O_2$ of which the members are:

Tiglic acid $C_5H_8O_2$
Oleic acid $C_{18}H_{34}O_2$
Elaïdic acid $C_{18}H_{34}O_2$
Iso-oleic acid $C_{18}H_{34}O_2$
Erucic acid $C_{22}H_{42}O_2$
Brassidic acid $C_{22}H_{42}O_2$

Of these, oleic acid (as glyceride) is the most widely distributed. Yet other series are:

The linolic $C_nH_{2n-4}O_2$
The linolenic $C_nH_{2n-6}O_2$
The clupanodonic $C_nH_{2n-8}O_2$
The ricinoleic $C_nH_{2n-2}O_3$

The fat which occurs in an oil-containing seed is not composed of the glyceride of one acid, but is a mixture of the glycerides of several, or even a large number of different acids, often members from more than one of the above series. Thus the fat of the fruit of the Coconut (*Cocos nucifera*) consists of a mixture of the glycerides of caproic, caprylic, capric, lauric, myristic, palmitic and oleic acids. Linseed oil from the seeds of *Linum usitatissimum* again is a mixture of the glycerides of palmitic, myristic, oleic, linolic, linolenic and isolinolenic acids. Similar mixtures are found in other fruits and seeds.

Since glycerol is a trihydric alcohol, it would be possible for one or more of the three hydroxyls to react with the acid to form mono-, di- or tri-glycerides. All these cases occur and, sometimes, one hydroxyl is replaced by one acid, and another hydroxyl by a different acid.

When the distribution of fats among the flowering plants is considered, they are found to be more widely distributed than the botanist is generally led to suppose.

The following is a list of some of the plants especially rich in fats as reserve material in the fruits or seeds. It represents only a selection of the better known genera, since many other plants have fatty seeds. An approximate percentage of oil present in the fruit or seed is given.

Gramineae: Maize (*Zea Mays*) 4 %.

Palmaceae: Oil Palm (*Elaeis guinensis*) 62 %: Coconut Palm (*Cocos nucifera*) 65 %.

Juglandaceae: Walnut (*Juglans regia*) 52 %.

Betulaceae: Hazel (*Corylus Avellana*) 55 %.

Moraceae: Hemp (*Cannabis sativa*) 33 %.

Papaveraceae: Opium Poppy (*Papaver somniferum*) 47 %.

Cruciferae: Garden Cress (*Lepidium sativum*) 25 %: Black Mustard (*Sinapis nigra*) 20 %: White Mustard (*Sinapis alba*) 25 %: Colza (*Brassica rapa* var. *oleifera*) 33 %: Rape (*Brassica napus*) 42 %.

Rosaceae: Almond (*Prunus Amygdalus*) 42 %: Peach (*P. Persica*) 35 %: Cherry (*P. Cerasus*) 35 %; Plum (*P. domestica*) 27 %.

Linaceae: Flax (*Linum usitatissimum*) 20–40 %.

Euphorbiaceae: Castor-oil (*Ricinus communis*) 51 %.

Malvaceae: Cotton (*Gossypium herbaceum*) 24 %.

Sterculiaceae: Cocoa (*Theobroma Cacao*) 54 %.

Lecythidaceae: Brazil Nut (*Bertholletia excelsa*) 68 %.

Oleaceae: Olive (*Olea europaea*) 20–70 %: Ash (*Fraxinus excelsior*) 27 %.

Rubiaceae: Coffee (*Coffea arabica*) 12 %.

Cucurbitaceae: Pumpkin (*Cucurbita Pepo*) 41 %.

Compositae: Sunflower (*Helianthus annuus*) 38 %.

The conclusion must not be drawn from the above list that the seeds of the plants mentioned have exclusively fats as reserve materials. In many cases fat may be the chief reserve product, but in others it may be accompanied by either starch or protein or both.

Some of the best-known examples of fat-containing seeds which yield "oils" of great importance in commerce, medicine, etc., are *Ricinus* (castor oil), *Brassica* (colza oil), *Gossypium* (cotton-seed oil), *Cocos* (coconut oil), *Elaeis* (palm oil), *Olea* (olive oil).

In the plant the fats are present as globules in the cells of the fat-containing tissues.

Plant fats may vary from liquids, through soft solids, to wax-like solids which generally have low melting-points. They float upon water in which they are insoluble. They are soluble in ether, petrol ether, benzene, chloroform, carbon tetrachloride, carbon bisulphide, etc.: some are soluble in alcohol. With osmic acid fats give a black colour, and they turn red with Alkanet pigment which they take into solution.

Expt. 90. *Tests for fats.* Weigh out 50 gms. of Linseed (*Linum usitatissimum*) and grind in a coffee-mill. Put the linseed meal into a flask, cover with ether, cork and allow the mixture to stand for 2–12 hrs. Filter off the ether into a flask, fit with a condenser and distil off the ether over an electric heater. (If a heater is not available, distil from a water-bath of boiling water after the flame has been turned out.) When the bulk of the ether is distilled off, pour the residue into an evaporating dish on a water-bath and drive off the rest of the ether. With the residue make the following tests in test-tubes:

(*a*) Try the solubilities of the oil in water, petrol ether, alcohol and chloroform It is insoluble in water and alcohol, but soluble in petrol ether and chloroform.

(*b*) Add a little 1 % solution of osmic acid. A black colour is formed. (This reaction is employed for the detection of fat in histological sections.)

(*c*) Add to the oil a small piece of Alkanet (*Anchusa officinalis*) root, and warm gently on a water-bath. The oil will be coloured red. Divide the oil into two portions in test-tubes. To one add a little water, to the other alcohol. The coloured oil will rise to the surface of the water in one case, and sink below the alcohol in the other. The Alkanet pigment being insoluble in both water and alcohol, these liquids remain uncoloured.

Keep some of the linseed oil for Expt. 91.

It is well known that the hydrocarbons of the unsaturated ethylene series C_nH_{2n} will combine directly with the halogens, chlorine, bromine and iodine to give additive compounds, thus:

$$C_2H_4 + Br_2 = C_2H_4Br_2$$
ethylene bromide

The acids of this series also behave in the same way, and since many plant fats contain members of the series, the fats will also combine with the halogens.

Expt. 91. To show the presence of unsaturated groups in a fat. To a little of the linseed extract add bromine water. Note the disappearance of the bromine and the formation of a solid product.

One of the most important chemical reactions of fats is that known as saponification. When a fat is heated with an alkaline hydroxide the following reaction takes place:

$$C_{17}H_{35}CO \cdot O—CH_2$$
$$|$$
$$C_{17}H_{35}CO \cdot O—CH \quad +3KOH = 3C_{17}H_{35}COOK + CH_2OH \cdot CHOH \cdot CH_2OH$$
glycerol
$$C_{17}H_{35}CO \cdot O—CH_2$$
tristearin

The potassium salt, potassium stearate, of the fatty acid, stearic acid, is termed a soap. The ordinary soaps used for washing are mixtures of such alkali salts of the various fatty acids occurring in vegetable and animal fats, and are manufactured on a large scale by saponifying fats with alkali. The soaps are soluble in water, so that when a fat is heated with a solution of caustic alkali, the final product is a solution of soap, glycerol and excess of alkali. The soap is insoluble in saturated salt (sodium chloride) solution, and when such a solution is added to the saponified mixture, the soap separates out and rises to the surface of the liquid. This process is known as "salting out." If the saponified mixture is allowed to cool without salting out, it sets to a jelly-like substance. When caustic potash is used for saponification and the product is allowed to set, a "soft" soap is formed. Hard soaps are prepared by using caustic soda and salting out.

The properties of soaps in solution are important. When a soap goes into solution, hydrolysis takes place to a certain extent with the formation of free fatty acid and free alkali. The free fatty acid then forms an acid salt with the unhydrolyzed soap. This acid salt gives rise to an opalescent solution and lowers the surface tension of the water with the result that a lather is readily formed.

The property of soaps of lowering surface tension is the reason for their producing very stable emulsions when added to oil and water (see chapter on colloids, p. 12).

Expt. 92. Hydrolysis of fat with alkali. Take 12 Brazil nuts, the seeds of *Bertholletia* (Lecythidaceae). Crack the seed coats and pound the kernels in a mortar. Put the pounded nut in a flask, cover it with ether, and allow the mixture to stand for 2–12 hrs. Filter into a weighed or counterpoised flask and distil off the ether as in Expt. 90. Weigh the oil roughly and add 4–5 times its weight of alcoholic caustic soda (prepared by dissolving caustic soda in about twice its weight of water and mixing the solution with twice its volume of alcohol). Heat on a water-bath until no oil can be detected when a drop of the mixture is let fall into a beaker of water. Then add saturated sodium chloride solution. The soaps will rise to the surface. Allow the soaps to separate out for a time and then filter. Press the soap dry with filter-paper, and test a portion to see that it will make a lather. Neutralize the filtrate from the soap with hydrochloric acid and evaporate as nearly as possible to dryness on a water-bath. Extract the residue with alcohol and filter. Test the filtrate for glycerol by means of the following tests:

(a) To a little of the solution add a few drops of copper sulphate solution and then some sodium hydroxide. A blue solution is obtained owing to the fact that glycerol prevents the precipitation of cupric hydroxide.

(b) Treat about 5 c.c. of a 0·5 % solution of borax with sufficient of a 1 % solution of phenolphthalein to produce a well-marked red colour. Add some of the glycerol solution (which has first been made neutral by adding acid) drop by drop until the red colour just disappears. Boil the solution: the colour returns. The reaction is probably explained thus. Sodium borate is slightly hydrolyzed in solution and boric acid, being a weak acid, is only feebly ionized, and therefore the solution is alkaline. On adding glycerol, glyceroboric acid (which is a strong acid) is formed and so the reaction changes to acid. On heating, the glyceroboric acid is hydrolyzed to glycerol and boric acid, and the solution again becomes alkaline.

(c) Heat a drop or two with solid potassium hydrogen sulphate in a dry test-tube; the pungent odour of acrolein (acrylic aldehyde) should be noted:

$$C_3H_8O_3 = C_2H_3 \cdot CHO + 2H_2O.$$

In addition to Brazil nuts, the following material can also be used: endosperm of Coconut, ground linseed, almond kernels and shelled seeds of the Castor-oil plant (*Ricinus*): about 50 gms. should be taken in each case.

Expt. 93. *Reactions of soaps.* (a) Take some of the soap which has been filtered off and shake up with water in a test-tube. A lather should be formed. (b) Make a solution of a little of the soap in a test-tube and divide it into three parts. To each add respectively a little barium chloride, calcium chloride and lead acetate solutions. The insoluble barium, calcium and lead salts will be precipitated. (The curd which is formed in the case of soap and hard water is the insoluble calcium salt.) Thirdly, take the remainder of the soap and acidify it with dilute acid in an evaporating dish, and warm a little on a water-bath. The soap is decomposed and the fatty acids are set free and rise to the surface.

Expt. 94. *Reactions of fatty acids.* (a) Try the solubilities in ether and alcohol of the acids from the previous experiment. They are soluble. (b) Shake an alcoholic solution of the fatty acids with dilute bromine water. The colour of the bromine is discharged owing to the bromine forming additive compounds with the unsaturated acids.

The question of the metabolism of fats in the plant is a very complicated one and has not yet been satisfactorily investigated. All plants may have the power of synthesizing fats, and a great number, as we have seen, contain large stores of these compounds in the tissues of the embryo, or endosperm, or both. The point of interest is that of tracing the processes by which these fats are synthesized, and are again hydrolyzed and decomposed. The products of decomposition may serve for the synthesis of other more vital compounds as the embryo develops, and before it is able to synthesize the initial carbohydrates, and to absorb the salts requisite for general plant metabolism.

One fact seems fairly clear, namely that when fat-containing seeds germinate, an enzyme is present in the tissues which has the power of hydrolyzing fats with the formation of fatty acids and glycerol. Such enzymes are termed lipases.

The lipase which has been most investigated is that which occurs in the seeds of the Castor-oil plant (*Ricinus communis*). It has been shown that if the germinating seeds are crushed and allowed to autolyze (p. 20) in the presence of an antiseptic, the amount of fatty acid increases, whereas in a control experiment in which the enzyme has been destroyed by heat, no such increase takes place (Reynolds Green, 13, 14).

Investigation has shown the enzyme to be present also in the resting seed, but in an inactive condition as a so-called zymogen (Armstrong, 4, 5, 6, 7). The zymogen is considered to be a salt and, after acidification with weak acid, the salt is decomposed, and the enzyme becomes active. After the preliminary treatment with acid, however, the enzyme is most active in neutral solution. The effect of acid on the zymogen may be demonstrated by autolyzing the crushed seed with a little dilute acetic acid; the increase of acidity will be found to be much greater than in the case of a control experiment in which acid has not been added. (See Appendix, p. 188.)

It has been found very difficult to extract the enzyme from the resting seed. An active material can be obtained by digesting the residue, after extraction of the fat, with dilute acetic acid and finally washing with water. This material can then be used for testing the hydrolytic power of the enzyme on various fats.

There is little doubt that lipase catalyzes the synthesis of fats as well as the hydrolysis; the reaction, in fact, has been carried out to a certain extent *in vitro*.

Expt. 95. Demonstration of the existence of lipase in ungerminated Ricinus *seeds.*

A. Remove the testas from about two dozen *Ricinus* seeds and pound the kernels up in a mortar. Into three small flasks (*a*), (*b*) and (*c*), put the following:

(*a*) 2 gms. of pounded seed + 10 c.c. of water.

(*b*) 2 gms. of pounded seed + 10 c.c. of water + 2 c.c. of N/10 acetic acid.

(*c*) 2 gms. of pounded seed + 10 c.c. of water + 2 c.c. of N/10 acetic acid, and boil well.

Add a few drops of chloroform to all three flasks, plug them with cotton-wool, and allow them to incubate for 12 hours at 37° C. Then add 2 c.c. of N/10 acetic acid to flask (*a*), and 25 c.c. of alcohol to all three flasks. Titrate the fatty acids present with N/10 alkali, using phenolphthalein as an indicator. A greater amount of fat should be hydrolyzed in (*b*) than in (*a*), and also slightly more in (*a*) than in (*c*). The addition of alcohol checks the hydrolytic dissociation of the soap formed on titration.

B. Pound up about 15 gms. of *Ricinus* seeds which have been freed from their testas, and let the pounded mass stand with ether for 12 hrs. Then filter, wash with ether and dry the residue. Weigh out three lots, of 2 gms. each, of the fat-free meal and treat as follows:

(*a*) Grind up the 2 gms. of meal in a mortar with 16 c.c. of N/10 acetic acid (i.e. 8 c.c. of acid to 1 gm. of meal), and let it stand for about 15 minutes. Then wash well with water to free from acid, and transfer the residue to a small flask. Add 5 c.c. of castor oil, 2 c.c. of water and a few drops of chloroform.

(*b*) Treat the 2 gms. of meal as in (*a*), but, after washing, and before transferring to the flask, boil well with a little distilled water. Add 5 c.c. of oil, 2 c.c. of water and a few drops of chloroform.

(*c*) Put the 2 gms. of meal into the flask without treatment and then add 5 c.c. of oil, 2 c.c. of water and a few drops of chloroform.

Incubate all three flasks for 12 hours, and then titrate with N/10 caustic soda, after addition of alcohol as in *A*. A certain amount of acetic acid is always retained by the seed residue, and this is ascertained from the value for flask (*b*). Flask (*c*) will act as the control.

Another question to be considered is the mode of synthesis in the plant of the complex fatty acids which form the components of the fats. No conclusive work has been done in this direction, but many investigators have held the view that the fats arise from carbohydrates, notably the sugars. In fact, it has been shown that in *Paeonia* and *Ricinus*, as the seeds mature, carbohydrates disappear and fats are formed.

The sequence of events, however, in the synthesis of fatty acids from sugars is very obscure. If we examine the formulae, respectively, of a hexose:

$$CH_2OH \cdot CHOH \cdot CHOH \cdot CHOH \cdot CHOH \cdot CHO$$

and a fatty acid, e.g. myristic acid:

$$H_3C—CH_2 \cdot CH_2 \cdot CH_2 \cdot CH_2 \cdot CH_2 \cdot CH_2 \cdot CH_2 \cdot CH_2 \cdot CH_2 \cdot CH_2 \cdot CH_2 \cdot CH_2 \cdot COOH$$

it is seen that three main changes are concerned in the synthesis of such a fatty acid from sugar, i.e. reduction of the hydroxyl groups of the sugar, conversion of the aldehyde group into an acid group, and finally the condensation or linking together of chains of carbon atoms. An interesting fact in connexion with this point is that all naturally occurring fatty acids have a straight, and not a branched, carbon chain and also contain an even, and not an odd, number of carbon atoms. It has been suggested (Smedley, etc., 15–17) that acetaldehyde and a ketonic acid, pyruvic acid, may be formed from sugar. By condensation of aldehyde and acid, another aldehyde is formed with two more carbon atoms. By repetition of the process, with final reduction, fatty acids with straight chains are produced.

WAXES

Waxes differ from fats in that they are esters of fatty acids with alcohols of high molecular weight of the methane series in place of glycerol. Such alcohols are cetyl alcohol, $C_{16}H_{33}OH$, carnaübyl alcohol, $C_{24}H_{49}OH$, ceryl alcohol, $C_{26}H_{53}OH$, and melissyl (or myricyl) alcohol, $C_{30}H_{61}OH$, etc.

Waxes occur as a deposit on the leaves, fruits and stems of many plants: they constitute, for instance, the "bloom" on the Grape, the Plum and the leaves of *Aloe*, *Mesembryanthemum*, etc., though they rarely occur in sufficient quantity to be readily collected. Nevertheless, the waxes of various plants have been isolated and analysed. The following are well known since they occur in considerable amounts:

Carnaüba wax is produced by the leaves of a Brazilian Palm (*Copernicia cerifera*). The leaves are detached and beaten, and the particles of wax collected and melted. About 2000–4000 leaves produce 16 kilos of wax.

Palm wax is obtained from the stem of the Wax Palm (*Ceroxylon andicolum*), a native of the Andes, and Raphia wax from the leaves of another palm (*Raphia Ruffia*). Pisang wax is produced by the leaves of a variety of the Banana (*Musa Cera*).

Waxes from different plants contain mixtures of various esters, of which the component alcohols have been mentioned above. The most commonly occurring acids are myristic, lignoceric, carnaübic, cerotic and melissic acids (see p. 89).

Expt. 96. *Tests for wax.* Take some commercial carnaüba wax and make the following experiments:

(*a*) Warm a small piece with alcohol in a test-tube. It goes into solution and separates out on cooling as a white crystalline deposit. Examine the crystals under the microscope.

(*b*) Warm a small piece with ether. It is soluble.

(*c*) Heat a small piece of wax with solid potassium hydrogen sulphate in a test-tube. There is no smell of acroleïn, since glycerol is absent [see Expt. 92 (*c*)].

PHYTOSTEROLS OR PLANT STEROLS.

These substances are unsaturated monohydric alcohols of high molecular weight of which the structural formulae are unknown. They are probably present in all parts of plants but the members most fully investigated have chiefly been obtained from seeds. The sterols are always found accompanying vegetable fats, and this connection is accentuated by the fact that they are soluble in the solvents used in fat extraction. When the fat is saponified, the sterols remain unaltered and are said to form the "unsaponifiable residue" of fats.

Various sterols have been isolated from different plants: many are isomeric and a usual formula is $C_{27}H_{45}OH$. One of the best defined sterols is **sitosterol** which occurs in the grain of the Wheat (*Triticum vulgare*) and Rye (*Secale cereale*): also in seeds of the Flax (*Linum usitatissimum*) and the Calabar Bean (*Physostigma venenosum*).

Expt. 97. *Detection of sterol in the grain of the Wheat.* Weigh out 300 gms. of grains and grind them in a coffee mill. Add 350 c.c. of ether to the ground mass in a flask, and allow it to stand for 24 hrs. Filter the extract through a pad of asbestos or glass wool in a funnel. Then wash the residue with another 150 c.c. of ether and filter. The ether extract is then saponified with sodium ethylate which is prepared as follows. Weigh out 2 gms. of metallic sodium, cut it into small pieces and add it slowly to 20 c.c. of 96–98% alcohol. When it has dissolved, add the solution of sodium ethylate to the ether extract in a separating funnel, shake well and allow the mixture to stand for at least 24 hours. Saponification takes place in the cold, and soap separates out. Filter, and shake up the filtrate several times with water in a separating funnel to remove alkali. Then evaporate off the ether in an evaporating basin on a water-bath after turning out the flame. Dissolve the unsaponifiable residue in a small quantity of hot 96–98 $\%$ alcohol and cool. A crystalline deposit of sterol will separate out. Examine under the microscope and note the elongated six-sided plates. Make 5 c.c. of a chloroform solution of some of the unsaponifiable residue and test for sterols as follows:

(a) To 2 c.c. of the chloroform extract add 2 c.c. of concentrated sulphuric acid. The chloroform layer develops a reddish-yellow to blood-red colour according to the amount of sterol present. The sulphuric acid layer shows a very characteristic green fluorescence. Pipette off the chloroform into a basin; it shows a play of colours, blue, green and yellow due to absorption of water.

(b) To 2 c.c. of the chloroform extract add 20 drops of acetic anhydride and then concentrated sulphuric acid drop by drop. A violet-pink colour appears which later changes to blue and green.

LECITHINS.

These substances are probably present in all living cells. True (pure) lecithin can be isolated from the animal, but preparations from the plant have hitherto always been mixtures with other substances. Various plant lecithins with such impurities have been isolated from seeds of the Wheat (*Triticum vulgare*), Castor-oil Plant (*Ricinus communis*), Pea (*Pisum sativum*), Lupin (*Lupinus*) and others: also from leaves of the Horse Chestnut (*Aesculus Hippocastanum*) and root of the Carrot (*Daucus Carota*).

Lecithin is a complex substance in which one hydroxyl of the glycerol of a fat forms an ester with phosphoric acid, the latter being also combined with the base, choline (see p. 170).

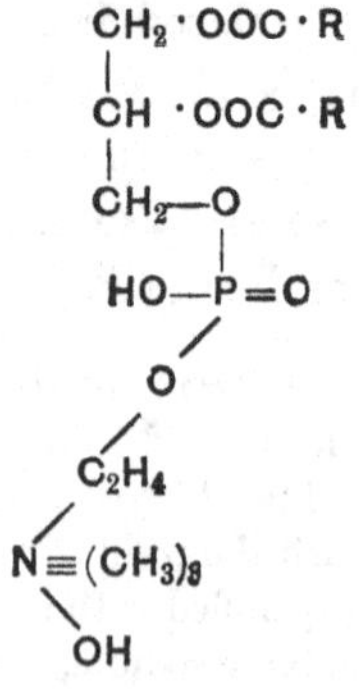

$$CH_2 \cdot OOC \cdot R$$
$$CH \cdot OOC \cdot R$$
$$CH_2—O$$
$$HO—P{=}O$$
$$O$$
$$C_2H_4$$
$$N{\equiv}(CH_3)_3$$
$$OH$$

Lecithins are yellowish wax-like substances which, on exposure to air, rapidly darken and become brown. They are hydrolysed by boiling with alkalies with the production of glycero-phosphoric acid, fatty acids and choline. The same decomposition is effected by lipase. An enzyme, **glycerophosphatase,** which decomposes glycero-phosphoric acid into phosphoric acid and glycerol has been shown to be present in bran and the seed of the Castor-oil Plant (*Ricinus communis*). Unlike lipase it is soluble in water (Plimmer, 12).

Expt. 98. *Tests for lecithin.* With commercial lecithin make the following tests:

(*a*) Test its solubility in ether, chloroform, benzene and carbon disulphide. It is soluble in all these solvents. To the ether solution add acetone; the lecithin is precipitated.

(*b*) Boil a little lecithin with alcohol in a test-tube. It is soluble.

(*c*) To the alcoholic solution from (*b*), add an alcoholic solution of cadmium chloride. A white precipitate of a double salt of lecithin and cadmium chloride separates out. Filter this off and test its solubilities in chloroform, benzene, etc. It is soluble. The double cadmium salt has been employed in the preparation and purification of lecithin.

(*d*) Heat a little lecithin with some strong caustic soda solution in a test-tube. Trimethylamine is evolved which can be detected by its smell. Acidify, and the fatty acids will separate out.

(*e*) Test for phosphoric acid in the following way. Weigh out 0·1 gm. of lecithin and mix it well with 1·4 gm. of potassium nitrate and 0·6 gm. of sodium carbonate. Incinerate the mixture in a porcelain crucible until it is colourless. Then dissolve the residue in the minimum amount of hot water, neutralize with hydrochloric acid, acidify with a few drops of concentrated nitric acid and pour the solution into an equal volume of boiling 3% ammonium molybdate solution. A yellow precipitate of ammonium phosphomolybdate is produced.

REFERENCES

BOOKS

1. **Abderhalden, E.** Biochemisches Handlexikon, III. Berlin, 1911.
2. **Allen's** Commercial Organic Analysis. Vol. 2. London, 1924. 5th ed.
3. **Leathes, J. B.,** and **Raper, H. S.** The Fats. London, 1925. 2nd ed.
4. **Lewkowitsch, J.** Chemical Technology and Analysis of Oils, Fats and Waxes. 6th ed. London, 1921.
5. **MacLean, H.,** and **MacLean, I. S.** Lecithin and allied Substances. The Lipins. London, 1927. 2nd ed.

PAPERS

6. **Armstrong, H. E.** Studies on Enzyme Action. Lipase. *Proc. R. Soc.,* 1905, B Vol. 76, pp. 606–608.
7. **Armstrong, H. E.,** and **Ormerod, E.** Studies on Enzyme Action. Lipase. II. *Proc. R. Soc.,* 1906, B Vol. 78, pp. 376–385.
8. **Armstrong, H. E.,** and **Gosney, H. W.** Studies on Enzyme Action. Lipase. III. *Proc. R. Soc.,* 1913, B Vol. 86, pp. 586–600.
9. **Armstrong, H. E.,** and **Gosney, H. W.** Studies on Enzyme Action. Lipase. IV. The Correlation of Synthetic and Hydrolytic Activity. *Proc. R. Soc.,* 1915, B Vol. 88, pp. 176–189.

10. **Ellis, M. T.** Contributions to our Knowledge of the Plant Sterols. Part I. The Sterol Content of Wheat (*Triticum sativum*). *Biochem. J.*, 1918, Vol. 12, pp. 160–172.

11. **Miller, E. C.** A Physiological Study of the Germination of *Helianthus annuus*. *Ann. Bot.*, 1910, Vol. 24, pp. 693–726. *Ibid.* 1912, Vol. 26, pp. 889–901.

12. **Plimmer, R. H. A.** The Metabolism of Organic Phosphorus Compounds. Their Hydrolysis by the Action of Enzymes. *Biochem. J.*, 1913, Vol. 7, pp. 43–71.

13. **Reynolds Green, J.** On the Germination of the Seed of the Castor-oil Plant (*Ricinus communis*). *Proc. R. Soc.*, 1890, Vol. 48, pp. 370–392.

14. **Reynolds Green, J.,** and **Jackson, H.** Further Observations on the Germination of the Seeds of the Castor-oil Plant (*Ricinus communis*). *Proc. R. Soc.*, 1906, B Vol. 77, pp. 69–85.

15. **Smedley, I.** The Biochemical Synthesis of Fatty Acids from Carbohydrate. *J. Physiol.*, 1912, Vol. 45, pp. xxv–xxvii.

16. **Smedley, I.,** and **Lubrzynska, E.** The Biochemical Synthesis of the Fatty Acids. *Biochem. J.*, 1913, Vol. 7, pp. 364–374.

17. **Lubrzynska, E.,** and **Smedley, I.** The Condensation of Aromatic Aldehydes with Pyruvic Acid. *Biochem. J.*, 1913, Vol. 7, pp. 375–379.

CHAPTER VIII

AROMATIC COMPOUNDS

THE aromatic compounds may be defined as substances containing the benzene carbon ring or a similar ring. A very great number occur among the higher plants but of these many are restricted in distribution, and may only be found in a few genera or even in one genus: others, on the other hand, are widely distributed. At present our knowledge of the part they play in general plant metabolism is slight.

The more widely distributed aromatic plant products may be grouped as:

1. The phenols, and their derivatives.
2. Inositol and phytin.
3. The aromatic acids, alcohols and aldehydes.
4. The tannins.
5. The "essential oils" and resins.
6. The flavone, flavonol and xanthone pigments, known as the soluble yellow colouring matters.
7. The anthocyan pigments, known as the soluble red, purple and blue colouring matters.

In connexion with the aromatic compounds it should be noted that many of them contain hydroxyl groups, and one or more of these groups may be replaced by the glucose residue, $C_6H_{11}O_5$—, with elimination of water and the formation of a glucoside, in the way already described (see p. 50). The majority of such compounds are sometimes classed together as a group—the glucosides—regardless of the special nature of the substance to which the glucose is attached (this course has been followed to some extent in Chapter X with compounds, the chief interest of which lies in their glucosidal nature). In treating of the aromatic substances in the following pages, mention will be made when they occur as glucosides, this combination being in these cases only a subsidiary point in their structure.

The various groups of aromatic substances will now be considered in detail.

PHENOLS.

There are three dihydroxy phenols, resorcinol, catechol and quinol, but only the two latter are known to exist in the free state in plants. Resorcinol frequently occurs as a constituent of complex plant products, and may be obtained on decomposition of such complexes by fusion with strong alkali, etc.

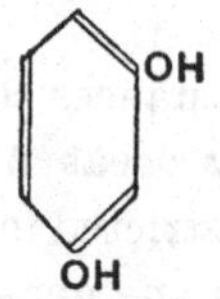

Resorcinol 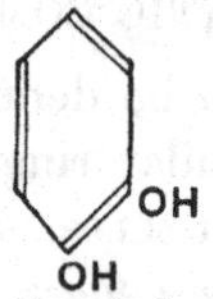   Catechol    Quinol

Quinol has been found in the free state in the leaves and flowers of the Cranberry (*Vaccinium Vitis-Idaea*). As a glucoside, known as arbutin, it occurs in many of the Ericaceae (see also p. 166).

Phloroglucinol is the only member of the trihydroxy phenols found uncombined in plants. It is very widely distributed in the combined state in various complex substances (Waage, 23).

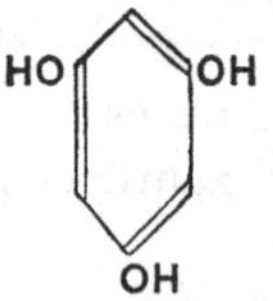

Phloroglucinol

INOSITOL AND PHYTIN.

Inositol is widely distributed in plants, especially in young leaves and growing shoots. It has been isolated from leaves of the Walnut (*Juglans regia*), fruit of the Mistletoe (*Viscum album*) and the unripe seed-pods of various plants. It is a polyhydric alcohol derived from benzene:

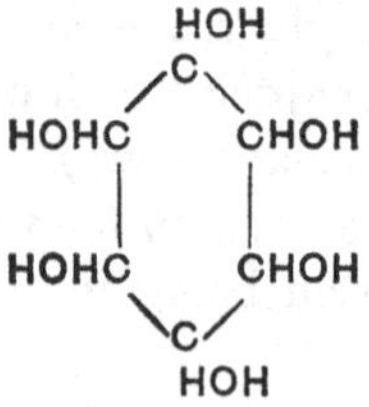

Inositol is soluble in water but crystallizes out on adding strong alcohol. It occurs also in seeds as the compound, **phytin**. The latter is an acid calcium and magnesium salt of inositol phosphoric acid which is a condensation product of inositol with six molecules of phosphoric acid (Plimmer and Page, 21). An enzyme, **phytase**, also occurring in seeds is able to hydrolyze phytin into inositol and phosphoric acid (Plimmer, 20).

AROMATIC ACIDS, ALDEHYDES AND ALCOHOLS.

There are two important series of these compounds found in the plant which can be represented respectively by benzoic acid and cinnamic acid and their derivatives:

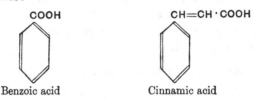

Benzoic acid   Cinnamic acid

Salicylic acid, or o-hydroxy-benzoic acid, occurs, both in the form of esters and in the free state, in various plants. The corresponding alcohol, **saligenin** or salicylic alcohol, in the form of the glucoside, salicin, occurs in the bark of certain species of Willow (*Salix*), and in the flower buds of the Meadow-sweet (*Spiraea Ulmaria*). Salicin is hydrolyzed by an enzyme contained in the plant in which it occurs into saligenin and glucose (see also p. 167). **Salicylic aldehyde** occurs in species of *Spiraea* and other plants.

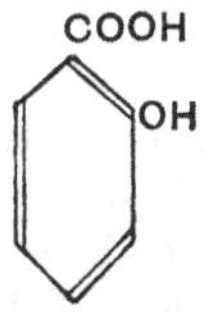

COOH   CH=CH·COOH

OH   OH

Salicylic acid 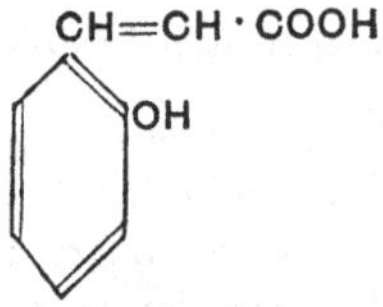 o-Coumaric acid

The corresponding derivative of cinnamic acid, i.e. o-**coumaric acid** is widely distributed as the anhydride, **coumarin** (see p. 165).

The relationship of cinnamic acid to phenylalanine and of p-coumaric acid to tyrosine (see p. 135) is important.

Protocatechuic acid is a dihydroxy-benzoic acid. It has been found in the free state in a few plants, but is more widely distributed as a constituent of many plant products. As will be shown later it forms the basis of the series of tannins.

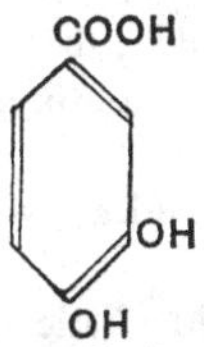

COOH   CH=CH·COOH

OH   OH

OH   OH

Protocatechuic acid 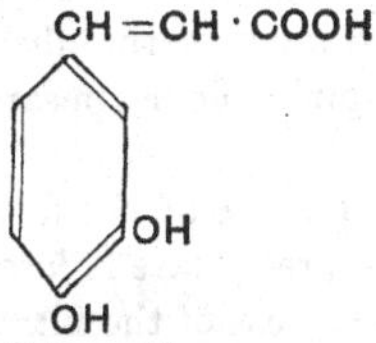   Caffeic acid

The corresponding derivative of cinnamic acid, i.e. **caffeic acid** (see also p. 123) is probably widely distributed. It is related to dihydroxy-phenylalanine (see p. 152).

Coniferyl alcohol is related to caffeic acid (see p. 103). Coniferyl alcohol, when oxidised, yields the aldehyde, vanillin (so much used for flavouring) which occurs in the fruits of the Orchid (*Vanilla plani-folia*). (See also p. 166.)

Gallic acid is a trihydroxy-benzoic acid:

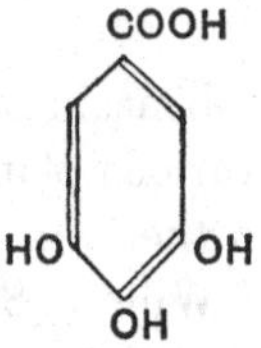

It occurs free in gall-nuts, in tea, wine, the bark of some trees and in various other plants. It forms a constituent of many tannins. It is a crystalline substance not very readily soluble in cold but more soluble in hot water. In alkaline solution it rapidly absorbs oxygen from the air and becomes brown in colour.

Expt. 99. The extraction and reactions of gallic acid. Take 100 gms. of tea, dry in a steam oven and grind in a mortar. Put the powder into a flask and cover well with ether. The preliminary drying and grinding can be omitted, but if carried out will make the extraction more complete. After at least 24 hrs. filter off the extract, and either distil or evaporate off the ether. The ether will be coloured deep green by the chlorophyll present in the dried leaves, and a green residue will be left. Add about 20 c.c. of distilled water to the residue, heat to boiling and filter. Heating is necessary because the gallic acid is only sparingly soluble in cold water. Keep the residue for Expt. 103. With the filtrate make the following tests; for (a), (b) and (c) dilute a few drops of the filtrate in a porcelain dish:

(a) Add a drop of 5 % ferric chloride solution. A blue-black coloration is given.

(b) Add a drop or two of iodine solution. A transient red colour appears.

(c) Add a drop or two of lime water. A reddish or blue coloration will be given.

(d) To a few c.c. of the filtrate in a porcelain dish add a little 5 % lead acetate solution. A precipitate is formed which turns red on addition of caustic potash solution, and dissolves to a red solution with excess of potash.

(e) To a few c.c. of the filtrate in a test-tube add a little 1 % potassium cyanide solution. A pink colour appears, but disappears on standing. On shaking with air it reappears.

(f) To a few c.c. of the filtrate in a test-tube add a few drops of 10 % gelatine solution. No precipitate is formed.

(g) To a few c.c. of the filtrate in a test-tube add a little 5 % lead nitrate solution. No precipitate is formed.

TANNINS.

This is a large group of substances, many of which are of complex composition. They arise in the plant from simpler compounds, such as protocatechuic, gallic and ellagic acids. Their formation takes place in various ways, either by condensation, accompanied by elimination of water, or by oxidation, or both; there may also be condensation with other aromatic complexes.

The tannins are widely distributed in the higher plants and, although no very systematic investigation has been made, it is obvious that some plants are rich in these substances, others poor, and others, again, apparently entirely without them. The tannins generally occur in solution in the cells of tissues of the root, stem, leaf, fruit, seed and flowers: sometimes they are confined to special cells, tannin-sacs, but after the death of the cell, the cell-walls of the dead tisssue become impregnated with the tannin. In tannin-producing plants, the tannin is generally found throughout the plant, and it probably tends to accumulate in permanent or dead tissues, such as the bark (dead cortex and cork), woody tissue, underground stems, etc.

Tannins appear to be more frequent in woody than in herbaceous plants, though in the latter they naturally only accumulate in the persistent underground stems and root-stocks. In annuals, also, tannins seem to be more rare: this may be due to the fact that in a short-lived plant, comparatively little tannin is formed and is not so readily detected as in the tissues of a perennial.

In certain plants which are highly tannin-producing and are also woody perennials, the bark becomes very rich in tannins. These barks are consequently of considerable commercial importance for tanning of leather. As examples may be taken species of *Caesalpinia*, Spanish Chestnut (*Castanea*), *Eucalyptus*, Oak (*Quercus*), Mangrove (*Rhizophora*), Sumac (*Rhus*). Tannins also occur in quantity in galls, especially on species of *Quercus*.

As a class, the tannins are non-crystalline and exist in the colloidal state in solution. They have a bitter astringent taste. They have certain properties and reactions in common, i.e. they precipitate gelatine from solution, are themselves precipitated from solution by potassium bichromate, and give either blue or green colorations with solutions of iron salts. Many tannins occur as glucosides but this is by no means always the case.

It is possible to classify the tannins into two groups according to whether they are complexes derived from protocatechuic acid or gallic acid:

1. The pyrogallol tannins. These give a dark blue colour with ferric chloride solution, and no precipitate with bromine water.

2. The catechol tannins. These give a greenish-black colour with iron salts, and a precipitate with bromine water.

Expt. 100. *Reactions of tannins.* Take three oak galls (the brown galls formed by species of *Cynips* on the Common Oak) and pound them finely in a mortar. Boil up the powder well with a small amount of water in an evaporating basin and let stand for a short time. Then filter. The filtrate will contain tannin together with impurities. Make the following tests with the extract:

(a) Put 2 c.c. of the tannin extract into a small evaporating dish, dilute with water, and add a drop or two of 5 % ferric chloride solution. A deep blue-black colour is produced.

(b) Put 2 or 3 drops of the tannin extract into a small evaporating dish, and dilute with water: add a little dilute ammonia and then a few drops of a dilute solution of potassium ferricyanide solution. A red coloration will appear.

(c) To 5 c.c. of the tannin solution in a test-tube add some strong potassium dichromate solution. The tannin will be precipitated.

(d) To about 5 c.c. of the tannin extract in a test-tube add a little 5 % lead acetate solution. The tannin will be precipitated.

(e) Melt a little of a 10 % solution of gelatine by warming gently and then pour drop by drop into a test-tube half full of tannin extract. The gelatine will be precipitated.

For the above tests, in addition to galls, the bark stripped from two or three year old twigs of *Quercus* may also be used, and will give the same reactions. The bark should be cut into small pieces for extraction.

It should be noted that although many tannins give the above reactions, it does not necessarily follow that all tannins will give all the reactions.

Expt. 101. *To demonstrate the existence of pyrogallol and catechol tannins.* The existence of a pyrogallol tannin which gives a blue reaction with iron salts has been illustrated in the last experiment on the Oak galls and the bark from Oak twigs. The bark of the Sumac (*Rhus Coriaria*) and the fruit pericarp, leaves and bark of the Sweet Chestnut (*Castanea vulgaris*) may be used as additional material for pyrogallol tannins.

For an iron-greening tannin strip off the outer bark from two to three year old twigs of the Horse Chestnut (*Aesculus Hippocastanum*). Cut or tear the bark into small pieces and boil *well* with a little water in an evaporating dish. Filter and test the filtrate with ferric chloride solution as in Expt. 101. A green coloration will be given. Iron-greening tannins may also be extracted from the bark of twigs of the Walnut (*Juglans regia*) and of the Larch (*Larix europaea*).

In the case of both classes of tannins, in addition to the ferric chloride reaction, the tests of Expt. 101 (c) and (e) should also be made on the extracts, in order to confirm the presence of tannin, since other substances, such as flavones, may give a green colour with iron salts (see p. 111).

Some of the individual tannins will now be considered.

Gallotannic (or tannic) acid is one of the most important of the pyrogallol tannins. It occurs in Oak galls and Oak wood, in tea, in the Sumac (*Rhus Coriaria*), etc. According to recent investigations (Fischer and Freudenburg, 8) tannic acid may be regarded as a compound of one molecule of glucose with five molecules of digallic acid in which five hydroxyls of the sugar are esterified by five molecules of acid:

$$CH_2(OX) \cdot CH \cdot CH(OX) \cdot CH(OX) \cdot CH(OX) \cdot CH(OX)$$
$$\underline{\qquad\qquad O \qquad\qquad}$$

where

$$X = -CO \cdot C_6H_2(OH)_2 \cdot O \cdot CO \cdot C_6H_2(OH)_3$$

Tannic acid is an almost colourless amorphous substance. It has an astringent taste, is soluble in water and alcohol, only slightly soluble in ether, and insoluble in chloroform. It is decomposed, by boiling with $2\,^0/_0$ hydrochloric acid, into gallic acid.

Expt. 102. *Extraction and reactions of tannic (or gallotannic) acid.* By a crude method a solution of gallotannic acid can be obtained from tea. About 5 gms. of the residue, after the extraction with ether in Expt. 100, is again extracted with ether once or twice which will remove all but traces of gallic acid. Boil up the residue from ether with a little water and filter. With the filtrate make the following tests which differentiate between gallic and gallotannic acid:

(a) To about 10 c.c. add a little $10\,^0/_0$ gelatine. The gelatine is precipitated.

(b) To a little of the filtrate add a few drops of $5\,^0/_0$ lead nitrate solution. The tannic acid is precipitated.

The remaining tests are given in common with gallic acid. If the extract is too coloured, dilute with water.

(c) Dilute a few drops of the filtrate with water in a porcelain dish and add a drop of $5\,^0/_0$ ferric chloride solution. A blue-black colour is given.

(d) Dilute a few drops of the filtrate with water in a porcelain dish and add a drop or two of iodine solution. A transient red colour is formed.

(e) To a little of the filtrate in a test-tube add a few drops of $1\,^0/_0$ potassium cyanide solution. A reddish-brown colour is formed which changes to brown but becomes red again on shaking with air.

In addition to tannic acid, a great many other tannins are known, but their constitution is obscure.

Expt. 103. *To demonstrate that in tannin-containing plants the tannin may be also present in the leaves.* Take about two dozen leaves of the Common Oak (*Quercus Robur*) and pound them in a mortar. Then boil the crushed mass in an evaporating dish with a little water. Filter, and with the filtrate make the tests for tannin. Leaves of other trees also may be used, e.g. the Wig Tree (*Rhus Cotinus*), Sweet Chestnut (*Castanea vulgaris*).

Expt. 104. *To demonstrate that tannins may be present in herbaceous as well as woody plants.* Extract some leaves, as in the last experiment, of Scarlet Geranium (*Pelargonium zonale*) and test for tannin.

Expt. 105. *To demonstrate that tannins may be present in petals and fruits, in addition to other parts of the plant.* Extract and test for tannins as in the last experiment, using petals of *Pelargonium zonale*, Common Paeony (*Paeonia officinalis*) or Rose (any garden variety), inflorescence of Flowering Currant (*Ribes sanguineum*), flowers of Horse Chestnut (*Aesculus Hippocastanum*) or pericarp of Sweet Chestnut (*Castanea*).

The "Essential Oils" and Resins.

When plant tissues are suspended in water, a current of steam passed through the suspension, and the distillate collected, a mixture of volatile substances will be found in the distillate and these can be separated from the water by various methods. Such a mixture of organic volatile products constitutes an "essential oil." The classification is purely artificial, as the mixture is heterogeneous and contains substances of very different chemical constitution. Since, however, the majority of "oils" consist largely of aromatic compounds, they are included in the present chapter. In many cases the "essential oil" contains some product of commercial value. About two hundred and fifty plants, representing between fifty and sixty Natural Orders, provide definite "oils," most of which are prepared commercially.

The chemical substances found in "essential oils" can be broadly classed as follows (see also p. 82).

1. The terpenes, which are complex, unsaturated (usually aromatic) hydrocarbons frequently of the formula, $C_{10}H_{16}$, e.g. pinene, limonene, caryophyllene and phellandrene.

2. Alcohols derived from the terpenes, e.g. borneol, menthol, citronellol[1], geraniol[1] and linalol[1]; corresponding aldehydes, e.g. citronellal[1] and other aromatic aldehydes, e.g. cinnamic aldehyde.

3. Esters of the above alcohols, e.g. bornyl acetate, geranyl acetate, linalyl acetate and menthyl acetate; also esters of other aromatic acids, e.g. methyl salicylate.

4. Phenols of high molecular weight, e.g. thymol, carvacrol and eugenol.

The following provide some examples of "essential oils":

"Oil of turpentine," from species of *Pinus*, *Larix* and *Abies*, contains pinene.

---

[1] The compound is aliphatic.

" Lavender oil," from *Lavandula vera* (Labiatae), contains limonene, linalyl acetate, linalol and others.

" Peppermint oil," from *Mentha piperata* (Labiatae), contains menthol, menthyl acetate and others.

" Clove oil," from *Eugenia caryophyllata* (Myrtaceae), contains eugenol and caryophyllene.

" Cinnamon oil," from *Cinnamomum zeylanicum* (Lauraceae), contains cinnamic aldehyde, eugenol and phellandrene.

" Lemon oil," from *Citrus Limonum* (Rutaceae), contains limonene, citronellol and citral.

" Thyme oil," from *Thymus vulgaris* (Labiatae), contains thymol and carvacrol.

" Rose oil," from *Rosa centifolia* (Rosaceae), contains citronellol, geraniol and others.

Camphor is a ketone derived from a solid terpene, camphene. The former occurs in the Camphor Tree (*Cinnamomum Camphora*), a genus of the Lauraceae.

The resins are oxidation products of the terpenes. They are differentiated into balsams and hard resins. The former consist of resins dissolved in, or mixed with, liquid terpenes, e.g. Canada balsam and crude turpentine. Copal and dammar are examples of hard resins.

Expt. 106. *Preparation of "clove oil" from cloves* (Wester, see p. 10). Cloves are the dried flower-buds of *Eugenia caryophyllata* (Myrtaceae). Take 100 gms. of cloves, pound them in a mortar and put the mass into a two litre flask one third full of water. Pass a current of steam through the flask, and collect the distillate cooled by a water condenser. The "essential oil" of *Eugenia* consists chiefly of the phenol, eugenol, $C_6H_4(OH)(OCH_3)CH_2CH{=}CH_2$, together with small quantities of the terpene, caryophyllene. The latter distills over first, but cannot be isolated unless much larger quantities of material are used. The eugenol settles out as an "oil" at the bottom of the watery distillate. Continue the distillation for four hours, or more, till all the eugenol has distilled over. Then add 25 gms. of sodium chloride for each 100 c.c. of the distillate, and shake up the mixture in a separating funnel with small quantities of petrol ether until no more eugenol can be extracted. The petrol ether extract is then distilled on a water bath (after the flame has been removed) to 25 c.c. Then extract it three times with 20 c.c. of 5 % sodium hydroxide solution in a separating funnel, whereby the sodium salt of the eugenol is formed and passes into the alkaline solution turning it yellow. The petrol ether now contains only the small quantity of the hydrocarbon, caryophyllene. Traces of the latter are now removed from the alkaline phenolate by extracting again with 20 c.c. of petrol ether. Then add dilute sulphuric acid to the phenolate. The eugenol separates out as a milky suspension, which gradually collects together as a yellow "oil." Then neutralise again with sodium carbonate solution (which does not form a phenolate), and extract the eugenol with petrol ether. Distil off the ether, and the eugenol remains.

THE FLAVONE AND FLAVONOL PIGMENTS.

These yellow colouring matters are very widely distributed in the higher plants (Shibata, Nagai and Kishida, 22). They are derived from the mother substances, flavone and flavonol, the latter only differing from the former in having the hydrogen in the central γ-pyrone ring substituted by hydroxyl:

Flavone   Flavonol

The naturally occurring pigments, however, have additional hydrogen atoms replaced by hydroxyl groups, that is they are hydroxy-flavones and flavonols, and the various members differ among each other in the number and position of these hydroxyl groups. Some of the members are widely distributed, others less so. Quite often more than one representative is present in a plant.

The flavone and flavonol pigments are yellow crystalline substances, and as members of a class they have similar properties. They occur in the plant most frequently as glucosides, one or more of the hydroxyl groups being replaced by glucose, or, sometimes, by some other hexose, or pentose. In the condition of glucosides, they are much less coloured than in the free state, and, being present in the cell-sap in very dilute solution, they do not produce any colour effect, especially in tissues containing chlorophyll. Occasionally they give a yellow colour to tissues, as in the rather rare case of some yellow flowers (*Antirrhinum*) where colour is due to soluble yellow pigment.

In the glucosidal state, the flavone and flavonol pigments are, as a rule, readily soluble in water and alcohol, but not in ether. In the nonglucosidal state they are, as a rule, readily soluble in alcohol, somewhat soluble in ether, but soluble with difficulty in water.

The flavone and flavonol pigments can be easily detected in any tissue by the fact that they give an intense yellow colour with alkalies (Wheldale, 24). If plant tissues be held over ammonia vapour, they turn bright yellow, showing the presence of flavone or flavonol pigments: the colour disappears again on neutralization with acids. (The reaction is especially well seen in tissues free from chlorophyll, such as white flowers.) This reaction will be found to be almost universal, showing how wide is their distribution. With iron salts, solutions of the pigments give green or brown colorations. With lead, insoluble salts are formed. Several of the members are powerful yellow dyes, and hence some plants in which they occur, such as Ling (*Erica cinerea*), Dyer's Weld or Rocket (*Reseda luteola*), have been used for dyeing purposes. The value of these colouring matters as dyes has led to their chemical investigation, and as a result the constitution, etc., of the hydroxy-flavones and flavonols is well established.

Expt. 107. *Demonstration of the presence of flavone or flavonol pigments in tissues without chlorophyll.* Take flowers of any of the undermentioned species and put them in a flask with a few drops of ammonia. They will rapidly turn yellow owing to the formation of the intensely yellow salt of the flavone or flavonol pigments present in the cell-sap. If the flowers are next treated with acid the yellow colour will disappear.

Also make an extract of some of the flowers with a little boiling water. Filter, cool and add the following reagents:

(*a*) A little alkali. A yellow colour is produced.

(*b*) A little ferric chloride solution. Either a green or brown coloration is produced.

(*c*) A little basic lead acetate solution. A yellow precipitate of the lead salt of the flavone or flavonol pigment is formed.

The flowers of the following species can be used: Snowdrop (*Galanthus nivalis*), Narcissus (*Narcissus poeticus*), white variety of Lilac (*Syringa vulgaris*), Hawthorn (*Crataegus Oxyacantha*), White Lily (*Lilium candidum*), white var. of Phlox, double white Pink, white Stock (*Matthiola*) etc., etc., in fact almost any species with white flowers or a white variety.

Expt. 108. *Demonstration of the presence of flavone or flavonol pigments in tissues containing chlorophyll.* Make a hot water extract of the leaves of any of the undermentioned species. Make with it the same tests as in the previous experiment.

Almost any green leaf may be used, but the following are suggested: Snowdrop (*Galanthus nivalis*), Dock (*Rumex obtusifolius*), Goutweed (*Aegopodium Podagraria*), Dandelion (*Taraxacum officinale*), Violet (*Viola odorata*), Ribwort Plantain (*Plantago lanceolata*), Elder (*Sambucus nigra*).

The most important *flavone* pigments are apigenin, chrysin and luteolin.

Apigenin has not yet been found to be widely distributed. Its formula is:

It occurs in the Parsley (*Carum Petroselinum*) (Perkin, 12) and in the flowers of the ivory-white variety of Snapdragon (*Antirrhinum majus*) (Wheldale and Bassett, 25).

Expt. 109. *Extraction of apiin, the glucoside of apigenin, from the Parsley* (Carum Petroselinum). Take some Parsley leaves and boil in as little water as possible. Filter off the extract and make the following tests for apigenin:

(*a*) Add alkali. A lemon yellow coloration is given.

(*b*) Add basic lead acetate solution. A lemon yellow precipitate is formed.

(*c*) Add ferric chloride solution. A brown colour is produced.

(*d*) Add ferrous sulphate solution. A reddish-brown colour is produced.

Apiin frequently separates out in a gelatinous condition from aqueous and dilute alcoholic solutions.

Chrysin is a flavone occurring in the buds of various species of Poplar (*Populus*). It has the formula:

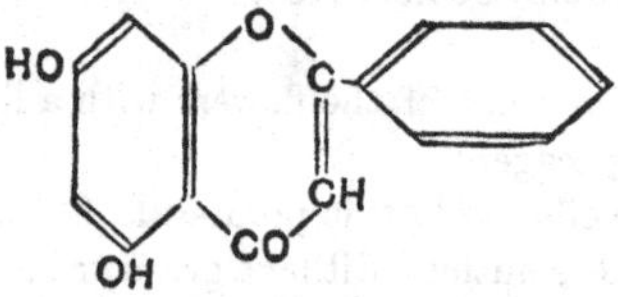

Luteolin does not appear to be widely distributed, though possibly it occurs in many plants in which it has not yet been demonstrated. Its formula is represented as:

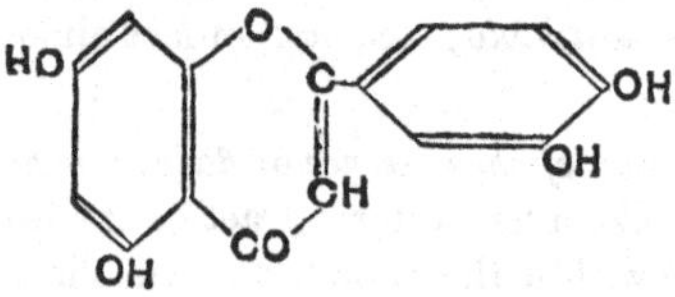

It occurs in the Dyer's Weld or Wild Mignonette (*Reseda luteola*) (Perkin, 11), Dyer's Greenweed or Broom (*Genista tinctoria*) (Perkin, 17) and in the yellow variety of flowers of the Snapdragon (*Antirrhinum majus*) (Wheldale and Bassett, 27). It has been much used as a yellow dye: hence the names of the first two plants (Perkin and Horsfall, 14).

The most important *flavonol* pigments are quercetin, kaempferol, myricetin and fisetin.

Quercetin is apparently one of the most widely distributed of the whole group of yellow pigments, and has the formula:

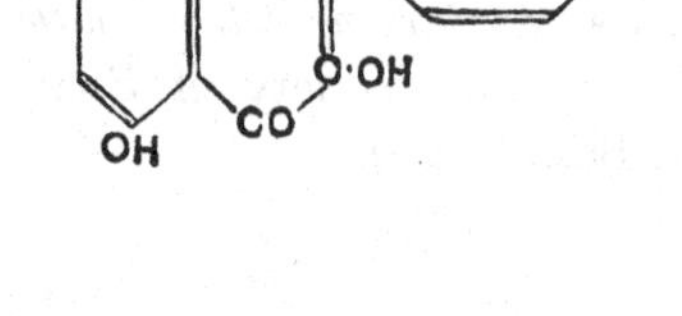

It occurs, either free, or combined with various sugars (glucose, rhamnose) as glucosides, in many plants, as for instance the following: in the bark of species of Oak (*Quercus*), in berries of species of Buckthorn (*Rhamnus*), in flowers of Wallflower (*Cheiranthus Cheiri*), Hawthorn (*Crataegus Oxyacantha*) (Perkin and Hummel, 16), Pansy (*Viola tricolor*) (Perkin, 13) and species of Narcissus: in leaves of Ling (*Calluna erica*) (Perkin, 17), and the outer scale leaves of Onion bulbs (Perkin and Hummel, 15).

Expt. 110. *Preparation of a glucoside of quercetin from flowers of either a species of Narcissus or the Wallflower* (Cheiranthus Cheiri). The most suitable species of *Narcissus* is *N. Tazetta*, but *N. incomparabilis* or any of the common yellow trumpet varieties such as the Daffodil (*N. Pseudo-Narcissus*) can be used. Take about 50 flowers of *Narcissus Tazetta* or about 20 gms. of petals of the Wallflower of either the brown or the yellow variety. The brown colour is due to a mixture of yellow plastid and of soluble purple (anthocyan) pigment in the sap. Pound the flowers in a mortar and then extract in a flask with boiling alcohol. Filter off the alcoholic extract and evaporate to dryness on a water-bath. Then add a little water and ether to the residue and transfer the whole to a separating funnel. The ether takes up the yellow plastid pigments, but the flavone and, in the case of the brown Wallflower, the anthocyan pigment remain in the water. Very soon, however, at the plane of separation of the liquids, the glucoside separates out as a crystalline deposit. This can be filtered off; with a dilute solution in alcohol make the following tests:

(*a*) Add a little alkali. The yellow colour is intensified, but the intensification disappears on adding acid.

(*b*) Add a little lead acetate solution. An orange precipitate of the lead salt is formed.

(*c*) Add a little ferric chloride solution. A green coloration is produced.

(*d*) Heat some of the alcoholic solution on a water-bath, acidify with strong hydrochloric acid and add zinc dust. A pink or magenta colour is produced (see p. 121).

Kaempferol occurs in the flowers of a species of Larkspur (*Delphinium consolida*) (Perkin and Wilkinson, 19) and *Prunus* (Perkin and Phipps, 18) and in the leaves or flowers of several other plants. It has the formula:

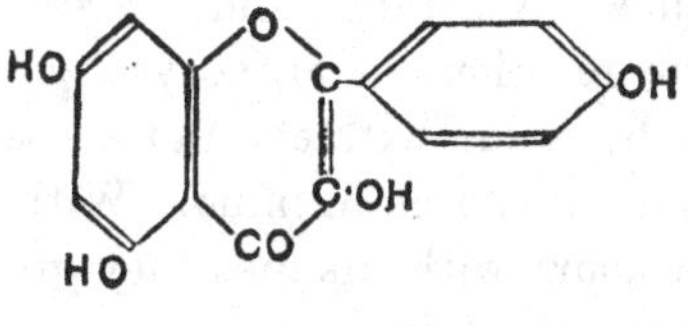

Myricetin and **fisetin** are two other flavones which have been found in species of Sumac (*Rhus*) and other plants. They have respectively the formulae:

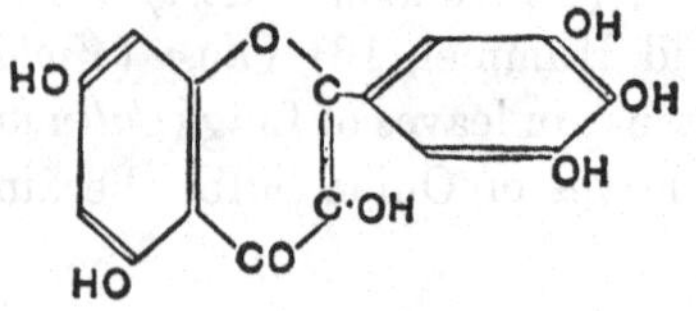

Myricetin 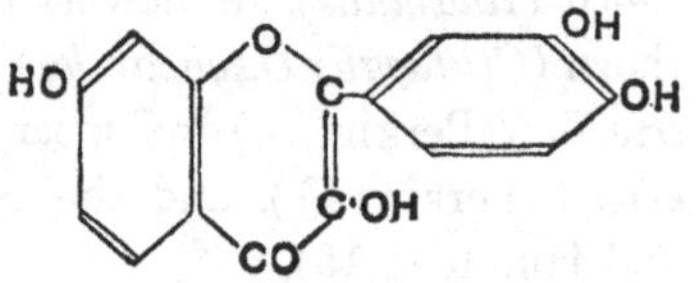   Fisetin

THE ANTHOCYAN PIGMENTS.

These pigments are the substances to which practically all the blue, purple and red colours of flowers, fruits, leaves and stems are due (Wheldale, 3). They occur in solution in the cell-sap and are very widely distributed, it being the exception to find a plant in which they are not produced. As members of a group, they have similar properties, but differ somewhat among themselves, the relationships between them being much the same as those between the various flavone and flavonol pigments. They occur in solution in the cell-sap but occasionally they crystallize out in the cell. They are present in the plant in the form of glucosides, and in this condition they are known as *anthocyanins*; as glucosides they are readily soluble in water and as a rule in alcohol [except blue Columbine (*Aquilegia*), Cornflower (*Centaurea Cyanus*) and some others] but are insoluble in ether and chloroform. The glucosides are hydrolyzed by boiling with dilute acids, and the resulting products, which are non-glucosidal, are termed *anthocyanidins* (Willstätter and Everest, 30). The latter, in the form of chlorides, are insoluble in ether, but are generally soluble in water and alcohol. The anthocyanins can be distinguished from the anthocyanidins in solution by the addition of amyl alcohol after acidification with sulphuric acid. The anthocyanidins pass over into the amyl alcohol, the anthocyanins do not. The anthocyanins and anthocyanidins themselves (with one exception) have not yet been crystallized, but of both classes crystalline derivatives with acids have been obtained (Willstätter and Everest, 30).

In considering the reactions of anthocyan pigments the difference between those given by crude extracts and those of the isolated and purified substances must be borne in mind. With acids the anthocyan pigments give a red colour: with alkalies they give, as a rule, a blue or violet colour when pure, but if flavone or flavonol pigments are present (as may be the case in a crude extract) they give a green colour, due to mixture of blue and yellow. In solution in neutral alcohol and water many anthocyan pigments lose colour, and this is said to be due to the conversion of the pigment into a colourless isomer which also gives a yellow colour with alkalies (Willstätter and Everest, 30); hence even a solution of a pure anthocyan pigment may give a green coloration with alkali due to mixture of blue and yellow. The isomerization can be prevented or lessened by addition of acids or of neutral salts which form protective addition compounds with the pigment. With lead acetate anthocyan pigments give insoluble lead salts, blue if the pigment is pure, or green, as in the case of alkalies, if it is mixed with flavone or flavonol pigments, or the colourless isomer.

When anthocyan pigments are treated with nascent hydrogen, the colour disappears but returns again on exposure to air. It is not known what reaction takes place.

Expt. 111. *The reactions of anthocyanins and anthocyanidins.* Extract petals of the plants mentioned below with boiling alcohol in a flask. Note that the anthocyan colour may disappear in the alcoholic extract. Filter off some of the alcoholic extract and make the following tests (*a*) and (*b*) with it :

(*a*) Add a little acid and note the bright red colour.

(*b*) Add a little alkali and note the green colour.

The decolorized petals, after filtering off the extract, should be warmed with a little water in an evaporating dish. The colour is brought back if pigment is still retained by them.

Evaporate the remainder of the alcoholic extract to dryness and note that the anthocyan colour returns. Dissolve the residue in water and continue the following tests, taking a little of the solution in each case :

(*c*) Add ether and shake. The anthocyan pigment is not soluble in ether.

(*d*) Add acid. A bright red colour is produced.

(*e*) Add alkali. A bluish-green or green colour is produced which may pass to yellow.

(*f*) Add basic or normal lead acetate solution. A bluish-green or green precipitate is produced.

(*g*) Add a little sulphuric acid and then amyl alcohol and shake ; the latter does not take up any of the red colour, indicating that the pigment is in the anthocyanin (glucosidal) state.

(*h*) Heat a little of the solution on a water-bath with dilute sulphuric acid and then cool and add amyl alcohol. The colour will pass into the amyl alcohol, indicating that the pigment is now in the anthocyanidin (non-glucosidal) state.

(*i*) Acidify a little of the solution with hydrochloric acid and add small quantities of zinc dust. The colour disappears. Filter off the solution and note that the colour rapidly returns again.

For the above reactions it is suggested that the following flowers be used as material : magenta Snapdragon (*Antirrhinum majus*), brown Wallflower (*Cheiranthus Cheiri*), crimson Paeony (*Paeonia officinalis*), magenta "Cabbage" Rose, Violet (*Viola odorata*), but the majority of coloured flowers will serve equally well.

Though the above represent the reactions and solubilities given by the greater number of anthocyan pigments, it will be found that all are not alike in these respects. Thus, for instance, the pigments of certain blue flowers, e.g. blue Larkspur (*Delphinium*), Cornflower (*Centaurea Cyanus*) and blue Columbine (*Aquilegia*) are neither soluble nor lose their colour in alcohol, but are soluble in water.

There is a small group of plants belonging to some allied natural orders, of which the anthocyan pigments give chemical reactions still more different from the general type already described, though they nevertheless resemble each other. Such, for instance, are the pigments of various genera of the Chenopodiaceae [Beet (*Beta*), Orache (*Atriplex*)], Amarantaceae (*Amaranthus* and other genera), Phytolaccaceae (*Phytolacca*) and Portulacaceae (*Portulaca*). These anthocyan pigments are insoluble in alcohol but soluble in water : they give a violet colour with acids, red to yellow with alkalies, and a red precipitate with basic lead acetate.

Anthocyan pigments may also occur in leaves, and this is very obvious in red-leaved varieties of various species such as the Copper Beech, the Red-leaved Hazel, etc.

Expt. 112. *Extraction of anthocyan pigment from the Red-leaved Hazel.* Extract some leaves of the Blood Hazel (*Corylus Avellana* var. *rubra*) with alcohol. Filter off and evaporate the solution to dryness. Add water. Pour a little of the crude mixture in the dish into a test-tube and add ether. There will be a separation into a green ethereal layer containing chlorophyll, and a lower water layer containing anthocyan pigment. Filter the extract remaining in the dish and with the filtrate make the tests already given in Expt. 111 (*c*)-(*i*).

The leaves of the Copper Beech (*Fagus sylvatica* var. *purpurea*) can also be used.

In many flowers, the cells of the corolla may contain, in addition to anthocyan, yellow plastid (see p. 40) pigments. The colour of the petals is in these cases the result of the combination of the two, and is usually some shade of brown, crimson or orange-red, as in the brown-flowered variety of Wallflower (*Cheiranthus Cheiri*), the bronze or crimson *Chrysanthemum*, the brown *Gaillardia* and the orange-red flowers of Nasturtium (*Tropaeolum majus*). The presence of the pigments can be demonstrated by their different solubilities (see Expt. 110).

Anthocyanins and anthocyanidins have been isolated from various species. The pigments themselves with one exception have not been obtained in the crystalline state, but crystalline compounds with acids have been prepared both of the glucosidal and non-glucosidal forms.

All the pigments so far described appear to be derived from three fundamental compounds, pelargonidin, cyanidin and delphinidin, of which the chlorides are represented thus:

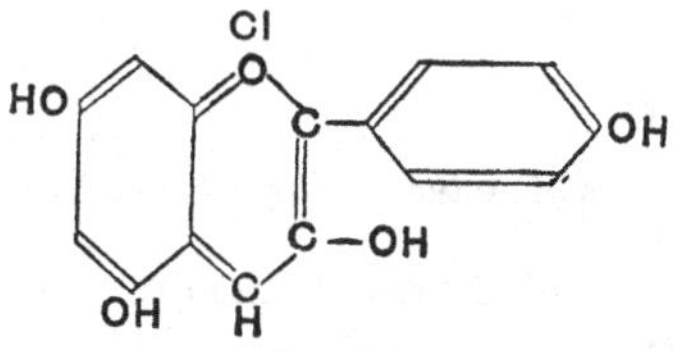

Pelargonidin chloride

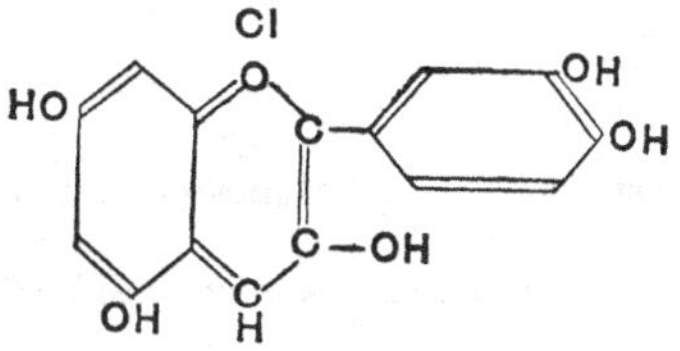

Cyanidin chloride

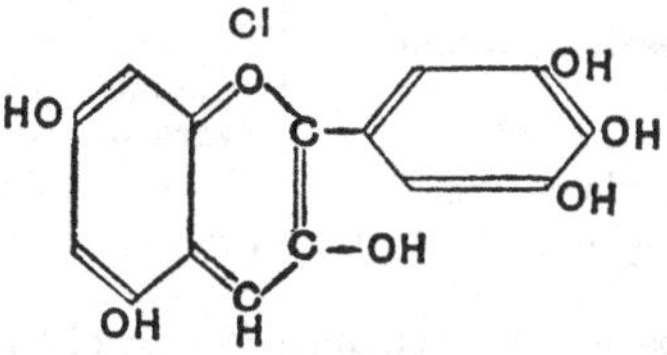

Delphinidin chloride

It has been suggested, at least in the case of cyanidin, the pigment of the Cornflower (*Centaurea Cyanus*), that the pigment itself is a neutral substance, purple in colour and of the following structure (Willstätter, 28, 31)·

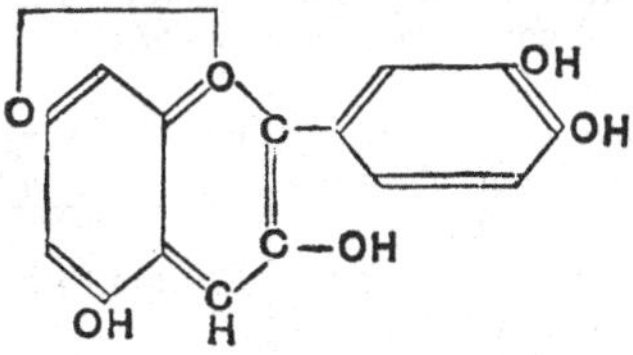

Further, that the blue pigment of the flower is the potassium salt of the purple, and the red acid salt, cyanidin chloride, depicted above, is a so-called oxonium compound of the purple.

Pelargonidin, moreover, has been prepared synthetically (Willstätter and Zechmeister, 33)

The above three pigments, either as glucosides or in the form of methylated derivatives, are found in a number of plants which are listed below (Willstätter, etc., 29, 32). The sugar residues or methyl groups may, of course, occupy different positions, thus giving rise to isomers:

Pelargonidin.

Callistephin	Monoglucoside of pelargonidin	Flowers of Aster (*Callistephus chinensis*)
Pelargonin	Diglucoside of pelargonidin	Flowers of Scarlet Geranium (*Pelargonium zonale*), pink var. of Cornflower (*Centaurea Cyanus*) and certain vars. of Dahlia (*D. variabilis*).

Cyanidin.

Asterin	Monoglucoside of cyanidin	Flowers of Aster (*Callistephus chinensis*)
Chrysanthemin	Monoglucoside of cyanidin	Flowers of Chrysanthemum (*C. indicum*)
Idaein	Monogalactoside of cyanidin	Fruit of Cranberry (*Vaccinium Vitis-Idaea*)
Cyanin	Diglucoside of cyanidin	Flowers of Cornflower (*Centaurea Cyanus*), *Rosa gallica* and certain vars. of Dahlia (*D. variabilis*)
Mekocyanin	Diglucoside of cyanidin	Flowers of Poppy (*Papaver Rhoeas*)
Keracyanin	Rhamnoglucoside of cyanidin	Fruit of Cherry (*Prunus Cerasus*)
Peonin	Diglucoside of peonidin (cyanidin monoethyl ether)	Flowers of Paeony (*Paeonia officinalis*)

Delphinidin.

Violanin	Rhamnoglucoside of delphinidin	Flowers of Pansy (*Viola tricolor*)
Delphinin	Diglucoside of delphinidin + *p*-hydroxybenzoic acid	Flowers of Larkspur (*Delphinium consolida*)
Ampelopsin	Monoglucoside of ampelopsidin (delphinidin monomethyl ether)	Fruit of Virginian Creeper (*Ampelopsis quinquefolia*)
Myrtillin	Monogalactoside of myrtillidin (delphinidin monomethyl ether)	Fruit of Bilberry (*Vaccinium Myrtillus*)
Althaein	Monoglucoside of myrtillidin	Flowers of deep purple var. of Hollyhock (*Althaea rosea*)
Petunin	Diglucoside of petunidin (delphinidin monomethyl ether)	Flowers of Petunia (*P. violacea*)
Malvin	Diglucoside of malvidin (delphinidin dimethyl ether)	Flowers of Mallow (*Malva sylvestris*)
Oenin	Monoglucoside of oenidin (delphinidin dimethyl ether)	Fruit of Grape (*Vitis vinifera*)

Of the methylated compounds, myrtillidin and oenidin may be represented thus:

Myrtillidin   Oenidin

Expt. 113. *Preparation and reactions of pelargonin chloride.* Extract the flowers from two or three large bosses of the Scarlet Geranium (*Pelargonium zonale*) in a flask with hot alcohol. Filter off and concentrate on a water-bath. Then pour the hot concentrated solution into about half its volume of strong hydrochloric acid. On cooling, a crystalline precipitate of pelargonin chloride separates out. Examine under the microscope and note that it consists of sheaves and rosettes of needles. Filter off the crystals, take up in water and make the following experiments with the solution:

(*a*) Add alkali. A deep blue-violet colour is produced.

(*b*) Take two equal quantities of solution in two evaporating dishes. To one add as quickly as possible some solid sodium chloride. The colour in the solution without salt will rapidly fade owing to the formation of the colourless isomer in neutral solution : this change is prevented to a considerable extent in the solution containing salt owing to the formation of an addition compound of the pelargonin with the sodium chloride which prevents isomerization (see p. 115). To portions of the water solution (without sodium chloride) which has lost its colour add respectively acid and alkali. The red colour returns with acid owing to the formation of the red acid oxonium salt : with alkali a greenish-yellow colour will be produced due to the formation of the salt of the colourless isomer. If alkali is added to the portion of the pigment solution containing the sodium chloride, it will be found that it still gives a violet colour.

(*c*) Add sulphuric acid and amyl alcohol. The alcohol does not take up the colour. Add amyl alcohol after acidifying another portion of the solution with sulphuric acid and heating on a water-bath. The alcohol now abstracts some of the colour. This shows that the glucoside pelargonin exists in the first case, but is decomposed into the non-glucosidal pelargonidin after heating with acid.

(*d*) Acidify with hydrochloric acid and add zinc dust : the colour disappears and returns again after filtering.

Expt. 114. *Preparation of the acetic acid salt of pelargonin.* Make an alcoholic extract of petals as in Expt. 113. Evaporate down and pour into glacial acetic acid instead of hydrochloric acid. The crystals of the salt formed are smaller and more purple in colour than those of the chloride.

In considering the anthocyan pigments, the question now arises— What is the chemical significance of the various shades in the living plant? Apparently the same pigment may be present in two flowers of totally different colours, as in the blue Cornflower and the magenta

Rosa gallica. It has been suggested that in such cases the pigment is modified by other substances present in the cell-sap: thus it may be present in one flower as a potassium salt, in another as an oxonium salt of an organic acid, and in a third in the unaltered condition. But exactly how these conditions are brought about is not clear. In one or two cases, moreover, where there is a red or pink variety of a blue or purple flower, the variety, when examined, has been found to contain a different pigment and one less highly oxidized than that in the species itself. The above phenomena are exemplified in the Cornflower (*Centaurea Cyanus*). The flowers of the blue type contain the potassium salt of cyanin, the purple variety, cyanin itself, while those of the pink variety contain pelargonin.

The mode of origin of anthocyan pigments in the plant is as yet obscure. It has been suggested (Wheldale, 24) that they have an intimate connexion with the flavone and flavonol pigments, which can be seen at once by comparing the structural formula of quercetin with that suggested for cyanidin:

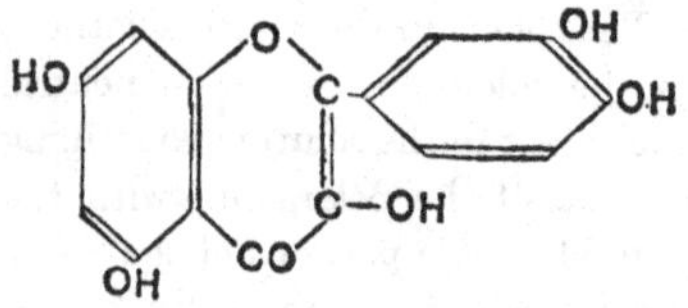

Quercetin 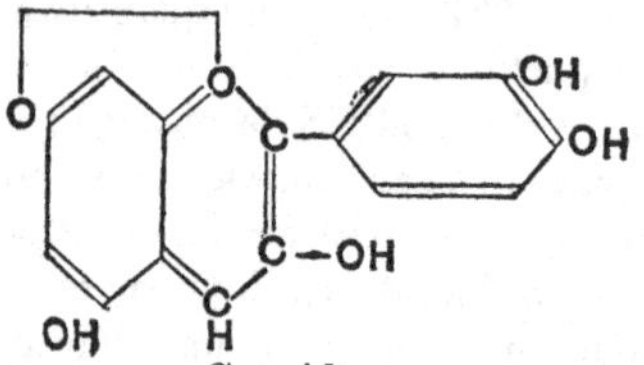   Cyanidin

All the anthocyan pigments so far isolated, however, have been found to contain the flavonol, and not the flavone, nucleus.

Just as in the case of the flavone and flavonol pigments, some of the anthocyan pigments are specific, while others, on the contrary, are common to various genera and species. Also more than one anthocyan pigment may be present in the same plant.

It will be pointed out later that small amounts of a substance identical with cyanidin are said to be formed by reduction of quercetin with nascent hydrogen, but this does not necessarily prove that the formation of anthocyan pigments in the plant takes place on the same lines. If we compare the formulae for a number of anthocyan with flavone and flavonol pigments, it is seen that they may be respectively arranged in a series, each member of which differs from the next by the addition of an atom of oxygen:

Luteolin, kaempferol and fisetin $C_{15}H_{10}O_6$ Pelargonidin $C_{15}H_{10}O_5$
Quercetin $C_{15}H_{10}O_7$ Cyanidin $C_{15}H_{10}O_6$
Myricetin $C_{15}H_{10}O_8$ Delphinidin $C_{15}H_{10}O_7$

The relationship between these two classes of substances in the plant can only be ascertained by discovering which flavone, flavonol and anthocyan pigments are present together[1], and then to determine whether the relationship is one of oxidation or reduction, a problem which has not yet received adequate attention (Everest, 7).

A reaction which is of interest in connexion with the relationship between the above two classes of pigments is that which takes place when solutions of some flavone or flavonol pigments are treated with nascent hydrogen. If an acid alcoholic solution of quercetin is treated with zinc dust, magnesium ribbon or sodium amalgam, a brilliant magenta or crimson solution is produced, and this solution gives a green colour with alkalies (Combes, 6). The red substance thus produced has been termed "artificial anthocyanin" or allocyanidin. The product is not a true anthocyan pigment but has, it is suggested, an open formation (Willstätter, 31):

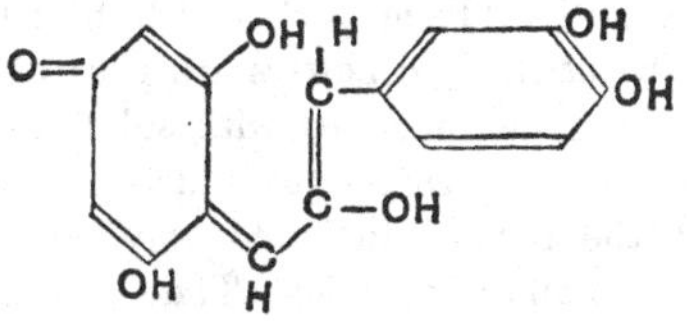

It is said, however, to contain small quantities of a substance identical with natural cyanidin from the Cornflower (Willstätter, 31). The fact that small quantities of a natural anthocyan pigment can be obtained artificially from a hydroxyflavonol by reduction does not necessarily imply that one class is derived from the other in the living plant.

From the above reaction of quercetin the result follows that when many plant extracts [most plants (see p. 110) contain flavone or flavonol pigments] are treated with nascent hydrogen, artificial anthocyan pigment is produced. Moreover, it seems probable that if the yellow pigments acted upon are in the glucosidal state, and if the reduction takes place in the cold, allocyanin (the glucoside of allocyanidin) is formed and the product is not extracted from solution by amyl alcohol. But if the flavone is non-glucosidal, or if the solution is boiled before or after reduction, then allocyanidin (non-glucosidal) is formed and is extracted by amyl alcohol.

Expt. 115. *Formation of allocyanidin from quercetin.* Make an alcoholic solution of a little of the glucoside of quercetin prepared from either *Narcissus* or *Cheiranthus*

---

[1] The only two satisfactory cases known are *Delphinium consolida,* which contains kaempferol and delphinidin, and *Viola tricolor*, which contains quercetin and delphinidin. Neither of these confirms the hypothesis of simple reduction.

(see Expt. 110). Acidify with a little strong hydrochloric acid and heat on a water-bath in an evaporating basin. Add a little zinc dust from time to time. A brilliant pink or magenta colour due to allocyanidin is produced. To a little of this solution add some alkali: a green colour is produced. If the alcohol and hydrochloric acid are evaporated off, and a little water and sulphuric acid added, on shaking up with amyl alcohol, all the allocyanidin passes into the amyl alcohol. (The distribution of the allocyanidin in the amyl alcohol is greater with aqueous sulphuric acid than with aqueous hydrochloric acid.)

Expt. 116. *Formation of allocyanin from quercetin.* Make a suspension of the glucoside of quercetin from *Cheiranthus* or *Narcissus* (see Expt. 110) in about 2N sulphuric acid, and then add zinc dust (or a drop of mercury about the size of a pea and a little magnesium powder) in the cold. The pink or magenta colour is gradually developed. Divide the coloured solution into two parts in two test tubes. Boil one for 5-10 minutes. Then add amyl alcohol to each. In the unboiled test-tube the amyl alcohol extracts no colour, since allocyanin is present. In the boiled test-tube allocyanidin is taken up by the amyl alcohol as in Expt. 115.

Expt. 117. *Formation of allocyanin and allocyanidin from plant extracts.* For this purpose the yellow varieties "Primrose" or "Cloth of Gold" of the Wallflower (*Cheiranthus Cheiri*) can be used. The flowers are pounded in a mortar, extracted with cold water, the water extract acidified with sulphuric acid, and zinc dust (or mercury and magnesium powder as above) added. A red coloration is slowly developed. To some of the red solution add amyl alcohol. The colour is not abstracted (allocyanin). Boil another portion. The allocyanin is thus converted into allocyanidin which is then taken up on addition of amyl alcohol.

OXIDIZING ENZYMES.

There are certain enzymes in the plant which are concerned with processes of oxidation and reduction (Chodat, 1). They are considered at this point since we have most information of them in their connexion with aromatic substances.

Peroxidases. A peroxidase is practically always present in the tissues of the Higher Plants. These enzymes are able to decompose hydrogen peroxide with the formation of "active" or atomic oxygen:

$$H_2O_2 + \text{peroxidase} = H_2O + O.$$

The tests for peroxidases will be considered later.

Oxidases (synonymous with laccases or phenolases) are only present in about 63 % of the Higher Plants. A plant oxidase, moreover, is made up of three components, i.e. (1) an enzyme, termed an oxygenase, (2) an aromatic substance containing an ortho-dihydroxy grouping such as that in catechol and (3) a peroxidase as above described (Wheldale Onslow, 9).

There are a number of substances with the catechol grouping, that is two hydroxyl groups in the ortho position, found in plants, such as catechol, protocatechuic acid, caffeic acid, hydrocaffeic acid, etc.,

Catechol   Protocatechuic acid   Caffeic acid

When solutions of such substances are left in air, they *slowly* autoxidize with the production of brownish oxidation products, accompanied, at the same time, by the formation of peroxide, probably hydrogen peroxide (since organic peroxides tend to decompose in the presence of water with the production of hydrogen peroxide). The oxygen in this form $-O-O-$ can be detected by chemical tests in the solutions. In plants, moreover, which contain catechol compounds, there are present certain enzymes, the oxygenases, which catalyze the autoxidation of the catechol compounds, and these only, with *rapid* production of a brown colour and of a peroxide[1]. Since peroxidases are also universally present, these may decompose the peroxide with production of active oxygen:

catechol substance + oxygenase + molecular oxygen⟶ peroxide

peroxide + peroxidase⟶ active oxygen.

This system, which constitutes an oxidase, is therefore capable of transforming molecular into active oxygen, and may in this way bring about oxidations in the plant. (See Appendix, p. 189.)

Catechol substances with the accompanying oxygenases are only present, as mentioned above, in about 63 % of the higher plants. They are present in about 76 % of the Monocotyledons, in about 84 % of the Sympetalae but only in about 50 % of the Archichlamydeae examined. Usually the genera of an order are all of one kind, either oxidase plants or peroxidase plants without the oxygenase and catechol elements. A few examples of oxidase orders are Gramineae, Umbelliferae, Labiatae, Boraginaceae, Solanaceae and Compositae: of peroxidase orders, Liliaceae, Cruciferae and Crassulaceae: of mixed orders, Ranunculaceae, Rosaceae and Leguminosae.

After death by injury, chloroform vapour, etc., the tissues of oxidase plants usually turn brown or reddish-brown in air, e.g. fruit of Apple, petals of *Anemone*, *Rosa*, etc.; peroxidase plants, on the contrary, do not

[1] The term oxygenase was originally applied by Bach and Chodat to ferment like compounds which form peroxides.

show this phenomenon. Since the oxidase provides an active oxidizing system, it is probable that a general oxidation of aromatic and other substances (in addition to catechol) takes place after death, in many cases leading to the production of dark pigments, e.g. the blackening of lacquer from latex of the Lacquer tree (*Rhus vernicifera*)[1]. In *Schenckia blumenaviana* (Rubiaceae), also, the whole plant turns bright red in chloroform vapour, and blue pigments are formed in flowers of an Orchid (*Phajus*) after death.

Tests for peroxidases are based on the property of a number of substances (benzidine, α-naphthol, guaiacum, pyrogallol, etc.) of giving highly coloured oxidation products in presence of active oxygen. Hence solutions of the above substances in the presence of hydrogen peroxide provide tests for peroxidases:

α-Naphthol 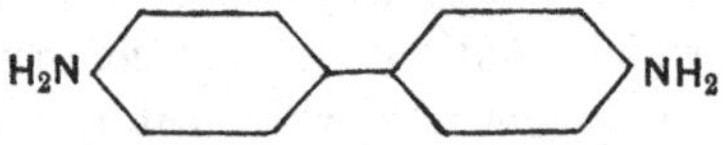   Benzidine 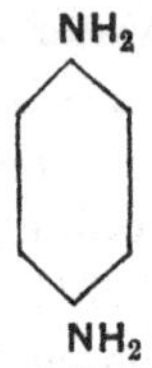   *p*-Phenylenediamine

Expt. 118. *Demonstration of the presence of a peroxidase.* Pound up a little Horse-radish root (*Cochlearia Armoracia*) with water. Filter and, taking a few c.c. each time in a small evaporating dish, make the following tests:

(*a*) Add a few drops of a 10 % solution of guaiacum. No colour is developed. Add a few drops of hydrogen peroxide: a deep blue colour appears.

Guaiacum gum is obtained from two West Indian species of Guaiacum trees, *G. officinale* and *G. sanctum*, partly as a natural exudation and partly by means of incisions. It gives a yellow solution with alcohol which contains guaiaconic acid, and the latter, on oxidation, yields guaiacum blue. As far as possible, inner portions of the resin lumps should be used, as the resin oxidizes in air, and then may give unreliable results. It is best to make the tincture freshly before use, and, as a precaution, to boil it on a water-bath with a little blood charcoal (preferably Merck's) and filter. Guaiacum gum tends to form peroxides on exposure to air, and these are removed by the above treatment.

(*b*) A 1 % solution of α-naphthol in 50 % alcohol, followed by a few drops of hydrogen peroxide. A lilac colour is developed.

(*c*) A 1 % solution of benzidine in 50 % alcohol followed by a few drops of hydrogen peroxide. A blue colour is developed.

(*d*) A 1 % solution of *p*-phenylenediamine hydrochloride in water followed by a few drops of hydrogen peroxide. A greenish colour is developed.

Repeat the above experiments with an enzyme extract that has been boiled. No colour is given, showing that the enzyme has been destroyed by boiling. Other

---

[1] The chief constituent of the latex, however, is a catechol derivative.

material which may be used for the above tests is fruit of the Melon and Cucumber and root of the Radish and Turnip.

Of the above substances only guaiacum, as a rule, is sufficiently sensitive to be oxidized by the amount of active oxygen produced by the plant oxidase. The juices and water extracts of oxidase plants will usually blue guaiacum immediately. If considerable quantities of sugars or tannins are present in the tissues, they may inhibit the guaiacum test.

Another test which may be used is the following. A solution of dimethyl-*p*-phenylenediamine hydrochloride and α-naphthol in presence of dilute sodium carbonate gives a deep violet-blue colour in the presence of an oxidase.

Expt. 119. *Demonstration of the presence of an oxidase.* Cut two or three thin slices from a fresh tuber of the Potato, pound *well* in a mortar, add a little water and filter. With a few c.c. of the extract in an evaporating dish make the following tests :

(*a*) Add a few drops of 10 % solution of guaiacum. A blue colour appears.

(*b*) Add 2·5 c.c. of a 0·14 % solution of α-naphthol and 2·5 c.c. of a 0·17 % solution of dimethyl-*p*-phenylenediamine hydrochloride and 5 c.c. of 0·1 % solution of sodium carbonate. A deep violet-blue colour appears.

Control experiments should be performed by using boiled enzyme extract. Other material which may be used is fruit of the Pear, Plum and Cherry.

Expt. 120. *To show the distribution of oxidases and peroxidases in various plants, and the correlation between the presence of oxidase and browning on injury or in chloroform vapour.* Take a selection of the plants given below, and in each case grind up a portion of the plant in a mortar with a little water and filter. Divide the filtrate into two parts in small porcelain dishes. Allow one part to stand in air, and note the darkening in colour in cases where an oxidase is present. To the other add a few drops of guaiacum. To extracts containing a peroxidase only, after 5–10 minutes, add in addition a few drops of hydrogen peroxide. Further, small pieces of the plants to be tested should be placed in a corked flask containing a few drops of chloroform, and the development of browning noted in the case of plants containing an oxidase. For demonstration of oxidases the following plants may be used : Christmas Rose (*Helleborus niger*), Dandelion (*Taraxacum officinale*), Forget-me-not (*Myosotis*), Hawthorn (*Crataegus*) and White Dead Nettle (*Lamium album*). For peroxidases : *Arabis, Aubrietia,* Pea (*Pisum sativum*), Stock (*Matthiola*), Wallflower (*Cheiranthus Cheiri*) and Violet (*Viola*).

The peroxidases, like other enzymes, can be extracted either with water or dilute alcohol and precipitated from solution by strong alcohol.

Expt. 121. *Preparation of peroxidase from Horse-radish* (Cochlearia) *roots.* Mince up the Horse-radish roots in a mincing machine. The product is allowed to stand for 24 hrs. to enable the glucoside, potassium myronate, to be hydrolyzed by the enzyme, myrosin. Then extract with 80 % alcohol. The alcohol is decanted off, and the

residue pressed free from alcohol in a press. The residue is next extracted with 40 %
alcohol for 48 hrs., filtered and precipitated with 90 % alcohol. The precipitate,
which contains the peroxidase, is filtered off. Dissolve up in water and make the
test for peroxidases (Expt. 118).

Peroxidase from the Horse-radish has been prepared on a large scale
and very carefully purified (Willstätter and Stoll, 34). The purified product was at first contaminated by a nitrogenous glucoside, which, however,
has been removed from subsequent preparations. (See Appendix, p. 192.)

The oxidation of pyrogallol, in the presence of a peroxidase and
hydrogen peroxide, has been used as a method for estimating the activity
of these enzymes. Solutions of known strength of pyrogallol and hydrogen peroxide are used, and to the mixture a solution of a known weight
of prepared peroxidase is added. An oxidation product, termed purpurogallin is formed. After a definite time, the reaction is stopped by adding
acid, and the purpurogallin extracted by ether. The ether extract is
colorimetrically compared with an extract containing a known amount
of purpurogallin (Willstätter and Stoll, 34).

Expt. 122. *Outline of method for estimating peroxidase by formation of purpurogallin.* Make a solution of 0·5 gm. of pyrogallol in 200 c.c. of distilled water, and
add to it 1 c.c. of 5 % hydrogen peroxide. Then add about 5 c.c. of a solution of
Horse-radish peroxidase from Expt. 121. After 5 minutes add to half the mixture
25 c.c. of dilute sulphuric acid and extract the purpurogallin with ether in a
separating funnel. The purpurogallin will be extracted by the ether, giving a yellow
solution. Allow the other half of the mixture to stand. The colour will deepen, and
a reddish deposit of purpurogallin will be precipitated. Examine a little of the
deposit under the microscope. It will be found to consist of sheaves of crystals.

A solution of peroxidase from *Alyssum* leaves [Expt. 124 (*b*)] can also be used.

The fact that an oxidase contains an oxygenase and catechol substance
may be demonstrated as follows. The tissue of an oxidase plant is rapidly
pounded under alcohol (to avoid oxidation) and extracted several times
with cold alcohol, by which the catechol substance is removed. The two
enzymes, oxygenase and peroxidase, remain in the tissue residue. This
residue or its water extract will give no (or very little) reaction with
guaiacum, since one of the components for producing the peroxide has
been removed. If now a little catechol is added followed by guaiacum,
a blue colour immediately appears. Moreover, from an alcoholic extract
of the tissues the catechol substance can be precipitated as a lead salt,
the lead removed as insoluble sulphate, and the aromatic compound set
free again in solution. If the enzyme extract is then added to the solution of the catechol substance, a brown colour is produced together with
peroxide, and the mixture will give a blue colour with guaiacum.

Expt. 123. *Resolution of the components of the oxidase in the Potato tuber.* (A) *Separation of peroxidase and oxygenase.* Cut a few thin slices from a peeled potato and put them in a mortar which contains sufficient 96 % alcohol to prevent, as far as possible, exposure to the air, and pound them thoroughly. Filter quickly on a filter-pump, and repeat the process several times until a colourless powder, consisting of cell-residues, starch, etc. is obtained. The enzymes (including the peroxidase and oxygenase) of the cells are precipitated by the alcohol and remain in the cell-residue. Make a water extract of the white powder and filter. To a portion of the filtrate add a few drops of guaiacum tincture; no blue colour is given. Add further a few drops of dilute hydrogen peroxide: a blue colour appears. (B) *Separation of the aromatic substance.* Take about 500 gms. of freshly peeled potato tuber, cut it into thin slices and drop them as rapidly as possible into a flask containing 250 c.c. of boiling 96 % alcohol on a water-bath. Continue boiling for 15 mins., and then filter. Evaporate off the alcohol from the filtrate, take up the residue in a little water, warm and filter. To the filtrate add concentrated lead acetate solution until a precipitate ceases to be formed. Filter off the precipitate, which is pale yellow in colour, stir up in a little water and add 10 % sulphuric acid drop by drop until the yellow colour is destroyed, and the lead is converted into lead sulphate. Filter off the lead sulphate: the filtrate contains the aromatic substance in solution. Neutralize the solution carefully with 1 % caustic soda and make the following tests with separate portions in small evaporating dishes:

(*a*) Add a drop of ferric chloride solution: a deep green colour appears. Add further a few drops of 1 % sodium carbonate solution. The green colour changes to a bluish- and finally, a reddish-purple. This reaction is characteristic of aromatic compounds containing the catechol grouping, i.e. two hydroxyl groups in the ortho position (see p. 123).

(*b*) Add a little of the enzyme solution prepared in (A). The mixture will gradually turn brown owing to the oxidation of the aromatic by the oxygenase.

(*c*) To (*b*) add a few drops of guaiacum tincture. A blue colour is given owing to the presence of the peroxide formed in (*b*), the oxidase system being now complete.

Expt. 124. *Action of oxygenase on catechol.* (*a*) *The oxygenase of the Potato tuber (or Pear fruit).* Make a 1 % solution of catechol in distilled water. To some of this solution, in a small evaporating dish, add a little of the enzyme solution from Expt. 123 (A). Note that the catechol solution gradually turns brown. Add further a few drops of guaiacum tincture. A blue colour appears. (*b*) *Enzyme extract of Alyssum leaves.* Pound up 2–3 *Alyssum* leaves in a mortar with some 96 % alcohol, and filter on a filter-pump. Repeat the process until the residue is practically colourless. Extract the residue with a little distilled water and filter. Proceed as in (*a*). No browning of catechol takes place and no blue colour is formed on the subsequent addition of guaiacum.

For section (*a*) the following material may also be used: fruits of Apple and Greengage, flowers of Horse Chestnut (*Aesculus*) and leaves of Pear, the method of preparation in (*b*) being employed. For section (*b*) flowers of white *Arabis* may also be used.

If in the preparation of the enzymes from the Potato tuber, the tissue is allowed to brown before extracting with alcohol, the cell-residue is tinged with brown and, on extraction with water, the filtrate will give an oxidase reaction with guaiacum. This is to be explained by the fact that oxidized products are adsorbed by the tissue residue. This pheno-menon is probably the explanation of the preparation of some oxidases called "laccases." Such enzymes have been obtained by the precipita-tion with strong alcohol of the juices (containing oxidized products since they were obtained by crushing the tissues) of plants which brown on injury. The enzyme and other organic matter is precipitated and carries with it the oxidized products which may readily oxidize phenols with other groupings. (See Appendix, p. 191.)

Tyrosinase. This enzyme is widely distributed in plants. It occurs in the Banana (*Musa sapientum*), Wheat (*Triticum vulgare*), Beet (*Beta vulgaris*), Oriental Poppy (*Papaver orientale*), Lacquer tree (*Rhus vernicifera*), Potato (*Solanum tuberosum*) and Dahlia (*Dahlia variabilis*). It has been demonstrated in about 16 natural orders and 21 genera.

Tyrosinase oxidizes tyrosine in a series of complicated reactions accompanied by the production of a pink colour which darkens through red to black. The final black pigments are known as melanins. A solu-tion of *p*-cresol.

can be used as a delicate test for the enzyme. If the enzyme is present, a yellowish or orange-red colour is formed.

Tyrosinase of the Potato tuber can be precipitated from a water extract with absolute alcohol: or if the potato tissue is extracted with cold 96 % alcohol, the enzyme is precipitated and remains in the tissue residue, as does the peroxidase (Expt. 123 (A)], but the tyrosine is almost entirely washed away.

Expt. 125. *Demonstration of the presence of tyrosinase in the Potato.* Take about half a potato and proceed as in the preparation of peroxidase [see Expt. 123 (A)]. Roughly dry the powder left on the filter and then add about 100 c.c. of water and allow to stand for 15 mins. Filter, and divide the filtrate into four portions *a*, *b*, *c* and *d*. Make a suspension of a little tyrosine in water (tyrosine is only slightly soluble in cold water).

To *a* add 5 c.c. of tyrosine suspension.
To *b* add 5 c.c. of tyrosine suspension and boil.
To *c* add some *p*-cresol.
To *d* nothing is added.

Plug all the tubes with cotton-wool, put in an incubator at 38° C. for 2-3 hrs. Note that tube *a* fairly rapidly turns red, then brown and finally black. Tube *d* may darken a little owing to the action of tyrosinase on traces of tyrosine or other plant aromatics left in the tissue. Tube *b* remains unaltered. Tube *c* gives an orange-red colour.

It is probable that tyrosinase is a mixture of enzymes, of which an oxidase is one component. It appears to be a fact that the plants which give the tyrosinase reactions are always oxidase, and not peroxidase, plants. (See Appendix, p. 191.)

Reductases. (Oxido-reductases.) These enzymes (Bach, 4) catalyze the decomposition of water into hydrogen and oxygen, provided substances are present which will accept the hydrogen and oxygen respectively. Such an enzyme has been shown to be present in the tuber of the Potato. It will reduce nitrates to nitrites, provided acetaldehyde is present, the latter being oxidized to acetic acid.

Expt. 126. *Demonstration of the presence of a reductase in the Potato.* Prepare a crude enzyme extract of the tuber as in [Expt. 123 (A)]. Take 10 c.c. of a 4 % solution of sodium nitrate in a test-tube, heat it in a beaker of water to 60° C. and then add 10 c.c. of the enzyme extract, followed by 3 drops of 10 % acetaldehyde solution. Prepare a control tube with boiled enzyme extract. Keep the tubes at 60° C. for 2-3 minutes. Test for nitrite with a few drops of an alcoholic solution of indole and a few drops of strong hydrochloric acid. The unboiled tube should give a red colour.

Catalases. These enzymes are probably present in all plants. They decompose hydrogen peroxide with the formation of molecular oxygen (see Expt. 15).

The function of the peroxidases, reductases, catalases and tyrosinase in the living cell is not known. It would appear that the oxidase reaction (as detected by guaiacum, etc.) is the outcome of post-mortem changes after the death of the cell. It is probable, however, that the processes giving rise to it may take place to some extent, though under control, in the living cell and it has been suggested, in fact, that oxidases play a part in respiration (Palladin, 10). There is certainly reason to believe that the first stages of respiration in plants involve a fermentation of a hexose similar to that taking place in yeast. The enzymes, **zymase** and **carboxylase** have been shown to be present in the tuber of the Potato and the root of Beet (Bodnar, 5). **Hexosephosphatase** has also been demonstrated in the bran of Wheat and seeds of the Castor-oil Plant (*Ricinus communis*) (Plimmer, 20). Whether oxidases act upon the products formed by the preliminary action of zymase remains an open question. The fact that they are not universally present in plants presents a difficulty.

REFERENCES

BOOKS

1. **Chodat, R.** Darstellung von Oxydasen und Katalasen tierischer und pflanz-licher Herkunft. Methoden ihrer Anwendung. Handbuch der biochemischen Arbeitsmethoden. E. Abderhalden, Berlin, 1910, Vol. 3 (1), pp. 42–74.

2. **Perkin, A. G.**, and **Everest, A. E.** The Natural Organic Colouring Matters. London, 1918.

3. **Wheldale, M.** The Anthocyanin Pigments of Plants. Cambridge, 1925. 2nd ed.

PAPERS

4. **Bach, A.** Zur Kenntnis der Reduktionsfermente. IV. Mitteilung. Pflanz-liche Perhydridase. *Biochem. Zs.*, 1913, Vol. 52, pp. 412–417.

5. **Bodnar, J.** Ueber die Zymase und Carboxylase der Kartoffel und Zuckerrübe. *Biochem. Zs.*, 1916, Vol. 73, 193–210.

6. **Combes, R.** Sur la présence, dans des feuilles et dans des fleurs ne formant pas d'anthocyane, de pigments jaunes pouvant être transformés en anthocyane. *C. R. Acad. sci.*, 1914, Vol. 158, pp. 272–274.

7. **Everest, A. E.** The Production of Anthocyanins and Anthocyanidins. Part III. *Proc. R. Soc.*, 1918, B Vol. 90, pp. 251–265.

8. **Fischer, E.**, und **Freudenberg, K.** Ueber das Tannin und die Synthese ähnlicher Stoffe. *Ber. D. chem. Ges.*, 1912, Vol. 45, pp. 915–935.

9. **Onslow, M. Wheldale.** Oxidising Enzymes. II. The Nature of the Enzymes associated with certain Direct Oxidising Systems in Plants. *Biochem. J.* 1920, Vol. 14, pp. 535–540. IV. The Distribution of Oxidising Enzymes among the Higher Plants. *Bioch. J.*, 1921, Vol. 15, pp. 107–112.

10. **Palladin, W.** Ueber das Wesen der Pflanzenatmung. *Biochem. Zs.*, 1909, Vol. 18, pp. 151–206.

11. **Perkin, A. G.** Luteolin. Part I. *J. Chem. Soc.*, 1896, Vol. 69, pp. 206–212. Part II. *Ibid.*, 1896, Vol. 69, pp. 799–803.

12. **Perkin. A. G.** Apiin and Apigenin. *J. Chem. Soc.*, 1897, Vol. 71, pp. 805–818. *Ibid.*, 1900, Vol. 77, pp. 416–423.

13. **Perkin, A. G.** Robinin, Violaquercetin, Myrticolorin and Osyritrin. *J. Chem. Soc.*, 1902, Vol. 81, pp. 473–480.

14. **Perkin, A. G.**, and **Horsfall, L. H.** Luteolin. Part III. *J. Chem. Soc.*, 1900, Vol. 77, pp. 1314–1324.

15. **Perkin, A. G.**, and **Hummel, J. J.** Occurrence of Quercetin in the Outer Skins of the Bulb of the Onion. *J. Chem. Soc.* 1896, Vol. 69, pp. 1295–1298.

16. **Perkin, A. G.**, and **Hummel, J. J.** The Colouring Matters occurring in various British Plants. Part I. *J. Chem. Soc.*, 1896, Vol. 69, pp. 1566–1572.

17. **Perkin, A. G.**, and **Newbury, F. G.** The Colouring Matters contained in Dyer's Broom (*Genista tinctoria*) and Heather (*Calluna vulgaris*). *J. Chem. Soc.*, 1899, Vol. 75, pp. 830–839.

18. **Perkin, A. G.**, and **Phipps, S.** Notes on some Natural Colouring Matters. *J. Chem. Soc.*, 1904, Vol. 85, pp. 56–64.

19. **Perkin, A. G.**, and **Wilkinson, E. J.** Colouring Matter from the Flowers of *Delphinium Consolida*. *J. Chem. Soc.*, 1902, Vol. 81, pp. 585–591.

20. **Plimmer, R. H. A.** The Metabolism of Organic Phosphorus Compounds Their Hydrolysis by the Action of Enzymes. *Biochem. J.*, 1913, Vol. 7, pp. 43–71.

21. **Plimmer, R. H. A.** and **Page, H. J.** An Investigation of Phytin. *Biochem. J.*, 1913, Vol. 7, pp. 157–174.

22. **Shibata, K., Nagai, I.,** and **Kishida, M.** The Occurrence and Physiological Significance of Flavone Derivatives in Plants. *J. Biol. Chem.*, 1916, Vol. 28, pp. 93–108.

23. **Waage, T.** Ueber das Vorkommen und die Rolle des Phloroglucins in der Pflanze. *Ber. D. bot. Ges.*, 1890, Vol. 8, pp. 250–292.

24. **Wheldale, M.** On the Nature of Anthocyanin. *Proc. Camb. Phil. Soc.*, 1909, Vol. 15, pp. 137–168.

25. **Wheldale, M.,** and **Bassett, H. Ll.** The Flower Pigments of *Antirrhinum majus*. II. The Pale Yellow or Ivory Pigment. *Biochem. J.*, 1913, Vol. 7, pp. 441–444.

26. **Wheldale, M.,** and **Bassett, H. Ll.** The Flower Pigments of *Antirrhinum majus*. III. The Red and Magenta Pigments. *Biochem. J.*, 1914, Vol. 8, pp. 204–208.

27. **Wheldale, M.,** and **Bassett, H. Ll.** The Chemical Interpretation of some Mendelian Factors for Flower-Colour. *Proc. R. Soc.*, 1914, B Vol. 87, pp. 300–311.

28. **Willstätter, R.** Ueber die Farbstoffe der Blüten und Früchte. *SitzBer. Ak. Wiss.*, 1914, pp. 402–411.

29. **Willstätter, R., Bolton, E. K., Mallison, H., Martin, K., Mieg, W., Nolan, T. S.,** und **Zollinger, E. H.** Untersuchungen über Anthocyane. Liebigs Ann. Chem., 1915, Vol. 408, pp. 1–162.

30. **Willstätter, R.,** und **Everest, A. E.** Ueber den Farbstoff der Kornblume. *Liebigs Ann. Chem.*, 1913, Vol. 401, pp. 189–232.

31. **Willstätter, R.,** und **Mallison, H.** Ueber die Verwandtschaft der Anthocyane und Flavone. *SitzBer. Ak. Wiss.*, 1914, pp. 769–777.

32. **Willstätter, R.,** und **Weil, F. J.** Untersuchungen über Anthocyane. *Liebigs Ann. Chem.* 1916, Vol. 412, pp. 113–251.

33. **Willstätter, R.,** und **Zechmeister, L.** Synthese des Pelargonidins. *SitzBer. Ak. Wiss.*, 1914, pp. 886–993.

34. **Willstätter, R.,** und **Stoll, A.** Ueber Peroxydase. *Liebigs Ann. Chem.*, 1918, Vol. 416, pp. 21–64.

CHAPTER IX

PROTEINS AND AMINO-ACIDS

No class of compounds is of more fundamental significance than the proteins. The matrix of protoplasm largely consists of proteins in the colloidal state, and, without doubt, they occur to some extent in the same condition in the cell-sap. They are also found in the cell in the solid state, in the form of either amorphous granules, termed aleurone, or crystalline or semi-crystalline bodies, termed crystalloids. Both solid forms constitute "reserve material" and are often found in seeds, tubers, bulbs, buds and roots.

Plant proteins may be classified on the following plan:

1. The simple proteins.
   (*a*) Albumins.
   (*b*) Globulins.
   (*c*) Prolamins (Gliadins).
   (*d*) Glutelins
2. Conjugated proteins.
   (*a*) Nucleoproteins.
3. Derived proteins.
   (*a*) Metaproteins.
   (*b*) Proteoses (Albumoses).
   (*c*) Peptones.
   (*d*) Polypeptides.

Although they are present in every cell in all parts of plants, little, however, is known of plant proteins, except of those in seeds, because of the difficulties of obtaining them in sufficiently large quantities, and of separating them from each other.

Proteins are in the colloidal state when in so-called solution, and are unable to diffuse through parchment membranes. The proteoses and peptones, however, which have simpler molecules, can diffuse through such membranes.

The vegetable proteins are soluble in various solvents according to the nature of the protein; some are soluble in water, others in dilute salt solutions, others, again, in dilute alkalies, and a few in dilute alcohol. Vegetable albumins are coagulated from solution on boiling, but most of the globulins, unlike the corresponding animal products, are only imperfectly coagulated on heating and some not at all. The precipitate formed when coagulation is complete will not go into solution again either in water, acid, alkali or salts. Alcohol precipitates the proteins; in the case of animal proteins, the precipitate becomes coagulated and insoluble if allowed to remain in contact with the alcohol but this does not appear to be so with plant proteins.

In addition, certain neutral salts, the chlorides and sulphates of sodium, magnesium and ammonium, have the property of precipitating proteins (except peptones) from solution when added in sufficient quantity. The protein is quite unchanged in precipitation and can be made to go into solution again. The various proteins are precipitated by different concentrations of these salt solutions (see p. 138).

The salts of calcium and barium and the heavy metals produce insoluble precipitates with the proteins, and in this case the reaction is irreversible.

In regard to chemical composition, the proteins contain the elements carbon, hydrogen, nitrogen, oxygen and sulphur. There is every reason to believe that the protein molecule is constituted of amino-acids condensed, with elimination of water, on the plan which may be depicted as follows:

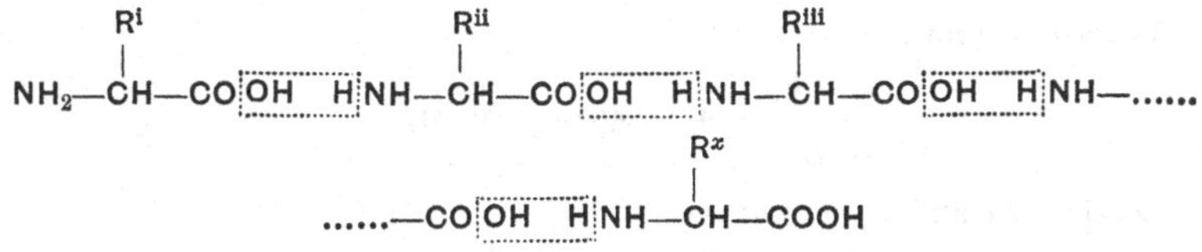

$$R^i \qquad\qquad R^{ii} \qquad\qquad R^{iii}$$
$$NH_2—CH—CO\overline{:OH \quad H:}NH—CH—CO\overline{:OH \quad H:}NH—CH—CO\overline{:OH \quad H:}NH—......$$
$$R^z$$
$$......—CO\overline{:OH \quad H:}NH—CH—COOH$$

Conversely, when the proteins are acted upon by hydrolyzing enzymes, a series of hydrolytic products are formed which have smaller molecules than the original proteins. They may be enumerated as:

1. Albumoses.
2. Peptones.
3. Amino-acids.

In the same way when proteins are boiled with acids, a number of the amino-acids are obtained as an end-product.

The above amino-acids may be either aliphatic or aromatic, and they are characterized by having one or more hydrogen atoms, other than those in the carboxyl groups, replaced by the group $-NH_2$. Thus they are acids by virtue of the carboxyl groups, and bases by virtue of the $-NH_2$ groups: towards strong acids they act as bases, and towards strong bases as acids. The amino-acid, alanine, for instance, forms salts, sodium amino-propionate with a base, and alanine hydrochloride with an acid:

$$\underset{\underset{NH_2}{|}}{CH_3—CH—COONa} \qquad\qquad \underset{\underset{NH_2 \cdot HCl}{|}}{CH_3—CH—COOH}$$

Substances behaving in this way have been termed "amphoteric" electrolytes (see also p. 16).

In the proteins, which are formed by condensation, as explained above, there are always some NH_2 and COOH groups left uncombined. Hence a protein must, in the same way, have the properties of both an acid and a base.

The amino-acids which are obtained by the hydrolysis of plant proteins may be classified as follows:

Aliphatic compounds.

Mono-carboxylic mono-amino acids :

  Glycine or a-amino-acetic acid
  $$CH_2(NH_2) \cdot COOH$$

  Alanine or a-amino-propionic acid
  $$CH_3 \cdot CH(NH_2) \cdot COOH$$

  Valine or a-amino-iso-valeric acid
  $$\underset{CH_3}{\overset{CH_3}{>}}CH \cdot CH(NH_2) \cdot COOH$$

  Leucine or a-amino-iso-caproic acid
  $$\underset{CH_3}{\overset{CH_3}{>}}CH \cdot CH_2 \cdot CH(NH_2) \cdot COOH$$

  Iso-leucine or a-amino-β-methyl-β-ethyl-propionic acid
  $$\underset{C_2H_5}{\overset{CH_3}{>}}CH \cdot CH(NH_2) \cdot COOH$$

  Serine or a-amino-β-hydroxy-propionic acid
  $$CH_2OH \cdot CH(NH_2) \cdot COOH$$

Dicarboxylic mono-amino acids :

  Aspartic acid or a-amino-succinic acid
  $$COOH \cdot CH_2 \cdot CH(NH_2) \cdot COOH$$

  Glutaminic acid or a-amino-glutaric acid
  $$COOH \cdot CH_2 \cdot CH_2 \cdot CH(NH_2) \cdot COOH$$

Mono-carboxylic di-amino acids :

Arginine or δ-guanidine-*a*-amino-valeric acid

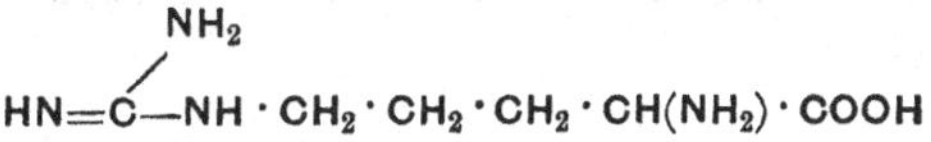

$$HN{=}\overset{\displaystyle NH_2}{C}{-}NH \cdot CH_2 \cdot CH_2 \cdot CH_2 \cdot CH(NH_2) \cdot COOH$$

Lysine or *a*-ε-di-amino-caproic acid

$$CH_2(NH_2) \cdot CH_2 \cdot CH_2 \cdot CH_2 \cdot CH(NH_2) \cdot COOH$$

Dicarboxylic di-amino acid :

Cystine (dicysteïne) or di-β-thio-*a*-amino-propionic acid

$$\begin{array}{cc} CH_2{-}S{-}S{-}CH_2 \\ | \qquad\qquad | \\ CH(NH_2) \qquad CH(NH_2) \\ | \qquad\qquad | \\ COOH \qquad\quad COOH \end{array}$$

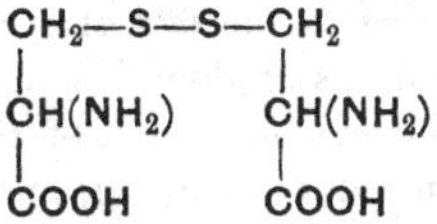

Aromatic compounds.

Mono-carboxylic mono-amino acids :

Phenyl-alanine or β-phenyl-*a*-amino-propionic acid

$$C_6H_5 \cdot CH_2 \cdot CH(NH_2) \cdot COOH$$

Tyrosine or *p*-hydroxy-phenyl-alanine

$$OH \cdot C_6H_4 \cdot CH_2 \cdot CH(NH_2) \cdot COOH$$

Heterocyclic compounds.

Proline or *a*-pyrrolidine-carboxylic acid

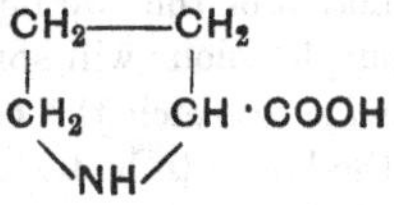

$$\begin{array}{c} CH_2{-}\!\!-CH_2 \\ | \qquad\quad | \\ CH_2 \quad CH \cdot COOH \\ \diagdown NH \diagup \end{array}$$

Histidine or β-iminazole-alanine

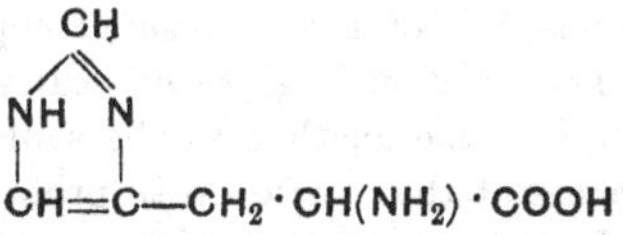

$$\begin{array}{c} CH \\ \diagup\!\!\diagdown \\ NH \quad N \\ | \qquad | \\ CH{=}C{-}CH_2 \cdot CH(NH_2) \cdot COOH \end{array}$$

Tryptophane or β-indole-alanine $C_8H_6N \cdot CH_2 \cdot CH (NH_2) \cdot COOH$

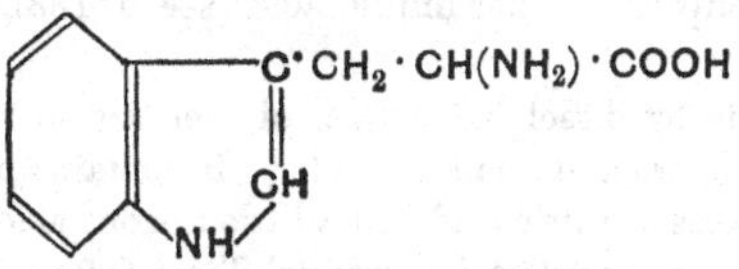

Different proteins are formed by various combinations of the above acids and hence give different amounts on hydrolysis.

There are certain properties and chemical reactions by means of which proteins can be detected. These are illustrated in the following experiment.

Expt. 127. *Tests for proteins.* Weigh out about 10 gms. of dried peas (*Pisum*), grind them in a coffee-mill and then add 100 c.c. of water to the ground mass. Allow the mixture to stand for an hour. Filter, and make the following tests with the filtrate (see p. 147).

(*a*) *The xanthroproteic reaction.* To a few c.c. of the protein solution in a test-tube add about one-third of its volume of strong nitric acid. A white precipitate is formed (except in the case of proteoses, peptones, etc.). On boiling, the precipitate turns yellow, and may partly dissolve to give a yellow solution. Cool under the tap, and add strong ammonia till the reaction is alkaline. The yellow colour becomes orange. The precipitate is due to the fact that metaprotein (see p. 143) is formed by the action of acid on albumins or globulins, and this metaprotein is insoluble in strong acids. The yellow colour is the result of the formation of a nitro-compound of some aromatic component of the protein, such as tyrosine, tryptophane and phenylalanine.

(*b*) *Millon's reaction.* To a few c.c. of the protein solution add about half its volume of Millon's reagent[1]. A white precipitate is formed. On warming, the precipitate turns brick-red, or disappears and gives a red solution. The white precipitate is due to the action of the mercuric nitrate on the proteins. The reaction is characteristic of all aromatic substances which contain a hydroxyl group attached to the benzene ring. The aromatic complex in the protein to which the reaction is due is tyrosine.

(*c*) *The glyoxylic reaction (Hopkins and Cole).* To about 2 c.c. of protein solution add an equal amount of "reduced oxalic acid[2]." Mix the solutions, and then add an equal volume of concentrated sulphuric acid, pouring it down the side of the tube. A purple ring forms at the junction of the two liquids. If the liquids are mixed by shaking the tube gently, the purple colour will spread throughout the solution. The substance in the protein molecule to which the reaction is due is tryptophane. If carbohydrates are present in the liquid to be tested, the colour is not good, owing to blackening produced by the charring with the strong sulphuric acid.

(*d*) *The biuret reaction.* To a few c.c. of the protein solution add about 1 c.c. of 40 % sodium hydrate and one drop of 1 % solution of copper sulphate. A violet or pink colour is produced. The reaction is given by nearly all substances containing two CONH groups attached to one another, to the same nitrogen atom, or to the same carbon atom. The cause of the reaction with proteins is the presence of one or more groupings formed by the condensation of the carboxylic group of an amino-acid with the amino group of another amino-acid (see p. 133).

[1] This reagent is made by dissolving 30 c.c. of mercury in 570 c.c. of concentrated nitric acid and then adding twice its bulk of water. It contains mercurous and mercuric nitrates, together with excess of nitric acid and a little nitrous acid.

[2] Reduced oxalic acid is prepared as follows: (*a*) Treat 500 c.c. of a saturated solution of oxalic acid with 40 gms. of 2 % sodium amalgam. When hydrogen ceases to be evolved, the solution is filtered and diluted with twice its volume of distilled water. The solution contains oxalic acid, sodium binoxalate and glyoxylic acid (COOH·CHO). (*b*) Put 10 gms. of powdered magnesium into a flask and just cover with distilled water. Add slowly 250 c.c. of saturated oxalic acid, cooling under the tap. Filter off the insoluble magnesium oxalate, acidify with acetic acid and dilute to a litre with distilled water.

(e) *The sulphur reaction.* Boil a few c.c. of the protein solution with an equal quantity of 40% sodium hydrate for two minutes, and then add a drop or two of lead acetate. The solution turns black (or brownish, if only a small amount of protein is present). This reaction is due to the formation of sodium sulphide by the action of the strong alkali on the sulphur of the protein. On addition of the lead salt, either a black precipitate, or dark colour, due to lead sulphide is formed. The sulphur in the protein molecule is mainly present as cystine.

For the following tests, a purified protein solution is necessary, since the reactions may also be given by accompanying aromatic substances, carbohydrates, etc. For this purpose take 40 gms. of ground peas, add to the meal about 200 c.c. 10% sodium chloride solution, and allow the mixture to stand, with occasional stirring, for 3–12 hrs. (see p. 147). Then filter off the extract, first through muslin, and, subsequently, through filter-paper. Put the extract to dialyze for 24 hrs. in a collodion dialyzer[1] until the protein is well precipitated. (Toluol should be added to the liquid in the dialyzer.) Then filter off the protein. Reserve half, and dissolve the other half in about 50 c.c. of 5% sodium nitrate solution. With this solution (after reserving a portion for Expt. 129) make the following tests:

(f) *Precipitation by alcohol.* To a few c.c. in a test-tube, add excess of absolute alcohol. The protein is precipitated.

(g) *Precipitation by the heavy metals.* Measure out a few c.c. of the protein solution into three test-tubes, and add respectively a little (1) 5% copper sulphate solution, (2) 5% lead acetate solution, (3) 5% mercuric chloride solution : the protein is precipitated in each case.

The following test cannot be demonstrated on the Pea protein, since carbohydrates are absent in this case. It can, however, be demonstrated in later experiments (see p. 145.

(h) *Molisch's reaction.* To a few c.c. of the protein solution add a few drops of a 1% solution of *a*-naphthol in alcohol. Mix, and then run in an equal volume of strong sulphuric acid down the side of the tube. A violet ring is formed at the junction of the two liquids. The reaction signifies the existence in a protein of a carbohydrate group which gives rise, on treatment with acid, to furfural. The latter then condenses with *a*-naphthol to give a purple colour (see also Expts. 39, 44, 46).

(i) *Precipitation by salts of alkaline earth metals.* To a few c.c. of the protein solution add a little 5% barium chloride solution. A precipitate is formed on standing.

(j) *Precipitation by neutral salts.* Saturate a few c.c. of the protein solution with finely powdered ammonium sulphate. The protein is precipitated or "salted out."

Since from a neutral salt solution the pea globulin is precipitated by acid (see p. 139), the tests (k)–(m) should be carried out with a solution of the protein in dilute acid. Dissolve, therefore, the remainder of the solid pea globulin in about 40 c.c. of 10% acetic acid, filter, and make the following tests:

(k) *Precipitation by tannic acid.* Add a little 3% tannic acid solution: the protein is precipitated.

---

[1] The collodion solution is made by adding 75 c.c. of ether to 3 gms. of well-dried pyroxylin, allowing it to stand for 10–15 minutes and then adding 25 c.c. of absolute alcohol. The dialyzers are prepared by coating the inside of a large test-tube with the solution and then filling with water, after the film is sufficiently dried so as not to be wrinkled by touching with the finger. The sac can then be withdrawn from the tube.

(*l*) *Precipitation by Esbach's solution*[1]. Add a little Esbach's solution: the protein is precipitated.

(*m*) *Precipitation by phosphotungstic acid.* Add a little 2% solution of phosphotungstic acid in 5% sulphuric acid: the protein is precipitated.

The substances used in the tests (*k*)–(*m*) are termed "alkaloidal reagents" because they also cause precipitation of alkaloids (see Chap. XI).

We are now in a position to deal with the different groups of proteins in detail:

SIMPLE PROTEINS.

Albumins. Very few vegetable albumins have been investigated. They can be best defined as proteins which are *soluble in water* and are coagulated by heat. Animal albumins are distinguished by the fact that they are not precipitated by saturating their neutral solutions with sodium chloride or magnesium sulphate; nor are they precipitated by half-saturation with ammonium sulphate. This distinction cannot be applied to vegetable proteins, since some are precipitated by the above treatment. It is often not easy to determine whether a plant protein is an albumin, on account of the difficulty of removing traces of salts, acids or bases which cause solubility, and also of separating the albumins from the globulins with which they occur. Albumins are however probably widely distributed in plant tissues.

The best-known albumins are:

Leucosin, which occurs in the seeds of Wheat (*Triticum vulgare*), Rye (*Secale cereale*) and Barley (*Hordeum vulgare*).

Legumelin, which occurs in seeds of the Pea (*Pisum sativum*), Broad Bean (*Vicia Faba*), Vetch (*Vicia sativa*), Lentil (*Ervum Lens*) and some other Leguminous seeds.

Phaselin, which occurs in the Kidney-bean (*Phaseolus vulgaris*).

Ricin, which occurs in the Castor-oil Bean (*Ricinus communis*).

Expt. 128. Demonstration of the presence of an albumin (leucosin) in wheat or barley flour (see also Expt. 135). Weigh out 10 gms. of wheat or barley flour, add 100 c.c. of distilled water and allow to stand, with occasional stirring, for 2–6 hrs. Then filter off the solution. Slowly heat the solution to boiling, and note that a precipitate of coagulated protein is formed.

Globulins. These may be defined as the proteins which are insoluble in water but soluble in dilute salt solutions, the concentration of the salt solution necessary for complete solution (see p. 139) varying with the salt or protein under consideration. It should be noted that,

[1] Esbach's solution is prepared by dissolving 10 gms. of picric acid and 10 gms. of citric acid in water and making the solution up to a litre.

in making *water*-extracts of plant tissues, it may happen that globulins pass into solution to some extent owing to the presence of inorganic salts in the tissues themselves. This has also already been illustrated in Expt. 127 in which an extract of the globulin of the Pea was obtained by treating ground Pea seeds with distilled water only.

It is characteristic of animal globulins that they are precipitated by saturation of their solutions with magnesium sulphate. Many of the vegetable globulins cannot be precipitated by the above means, though they are all, as far as tested, precipitated by sodium sulphate at 33° C. Many also (like animal globulins) are precipitated by half-saturation with ammonium sulphate, though others are not precipitated until their solutions are nearly saturated with this salt [see Expt. 127 (*j*)].

Unlike animal globulins, vegetable globulins are, as a rule, only imperfectly coagulated by heat, even on boiling.

Expt. 129. Demonstration of the coagulation of globulin. Heat a few c.c. of the solution of dialyzed Pea globulin (from Expt. 127) in a test-tube. Note that the protein is largely precipitated, but the solution does not become quite clear.

One very important characteristic of the vegetable globulins is the ease with which a number of them can be obtained in crystalline form. This result may be achieved by dialyzing a salt solution of the globulin. The salt passes out through the membrane, and the protein is deposited in the form of crystals. An alternative method is to dilute the saline solution of globulin with water at 50—80° C. until a slight turbidity appears. Then warm further until this goes into solution, and cool gradually, when the protein will separate in crystals. The globulin, edestin, from seeds of the Hemp (*Cannabis sativa*) crystallizes very readily (see Expt. 139) and crystals can also be obtained of the globulins from the seeds of the Brazil nut (*Bertholletia excelsa*), the Flax or Linseed (*Linum usitatissimum*), the Oat (*Avena sativa*) and the Castor-oil plant (*Ricinus communis*); other globulins separate out on dialysis as spheroids, sometimes mixed with crystals.

The solubilities of plant globulins are further complicated by the fact that some of these substances form acid salts which have different solubilities from the proteins themselves. Thus edestin is insoluble in water, but soluble in either dilute salt solution or acid. In the presence of acid it forms salts which are insoluble in dilute salt solutions. Thus edestin in dilute acid solution is precipitated by a trace of salt, or in dilute salt solution by a trace of acid (see Expt. 130). Legumin, on the other hand, from the Pea and other Leguminosae is soluble in water in the free state; combined with a small amount of acid as a salt, it is insoluble in water but soluble in neutral salt solution, that is, it has the solubilities of a globulin (see p. 147).

Expt. 130. *The formation of salts by edestin.* Grind up 5 gms. of seeds of the Hemp (*Cannabis sativa*) in a coffee-mill. Extract with 50 c.c. of warm (not above 60° C.) 10% sodium chloride solution and filter. Add a drop of strong hydrochloric acid to the filtrate. Edestin chloride, which is insoluble in salt solutions, is precipitated. Filter and drain off all the liquid, wash once and then suspend the precipitate in distilled water. Add 1 or 2 drops of hydrochloric acid carefully and stir till most or all of the precipitate goes into solution. Filter, and to the filtrate add a few drops of saturated sodium chloride solution. The edestin acid salt is again precipitated.

The following is a list of the principal known globulins (Osborne, 2):

Legumin, in seeds of	Pea (*Pisum sativum*). Broad Bean (*Vicia Faba*). Vetch (*Vicia sativa*). Lentil (*Ervum Lens*).
Vignin, in seeds of	Cow Pea (*Vigna sinensis*).
Glycinin, in seeds of	Soy Bean (*Glycine hispida*).
Phaseolin (crystalline), in seeds of	Kidney Bean (*Phaseolus vulgaris*). Adzuki Bean (*P. radiatus*). Lima Bean (*P. lunatus*).
Conglutin, in seeds of	Lupin (*Lupinus*).
Vicilin, in seeds of	Pea (*Pisum sativum*). Broad Bean (*Vicia Faba*). Lentil (*Ervum Lens*).
Corylin, in seeds of	Hazel Nut (*Corylus Avellana*).
Amandin, in seeds of	Almond (*Prunus Amygdalus*). Peach (*P. Persica*). Plum (*P. domestica*). Apricot (*P. Armeniaca*).
Juglansin, in seeds of..............	European Walnut (*Juglans regia*). American Black Walnut (*J. nigra*). American Butter-nut (*J. cinerea*).
Excelsin (crystalline), in seeds of	Brazil Nut (*Bertholletia excelsa*).
Edestin in seeds of.................	Hemp (*Cannabis sativa*).
Avenalin, in seeds of	Oat (*Avena sativa*).
Castanin, in seeds of	Sweet Chestnut(*Castanea vulgaris*).
Maysin, in seeds of.................	Maize (*Zea Mays*).
Tuberin, in tubers of	Potato (*Solanum tuberosum*).

Crystalline globulins have also been isolated from the following seeds but have as yet no distinctive names: Flax (*Linum usitatissimum*), Squash (*Cucurbita maxima*), Castor-oil Bean (*Ricinus communis*), Coconut (*Cocos nucifera*), Cotton-seed (*Gossypium herbaceum*), Sunflower (*Helianthus annuus*), Radish (*Raphanus sativus*), Peanut (*Arachis hypogaea*), Rape (*Brassica campestris*) and Mustard (*Brassica alba*).

It will be seen that the majority of reserve proteins of seeds are globulins. It is probable that native and artificial crystalline proteins are identical in many cases.

Prolamins. These proteins are characterized by the fact that they are insoluble in water and dilute saline solutions, but are soluble in 70–90 % alcohol. Such proteins are peculiar to plants, and are formed to a considerable extent in the seeds of cereals. The principal ones which have been isolated are:

Gliadin found in the seeds of Wheat (*Triticum vulgare*).

,, ,, ,, Rye (*Secale cereale*).

Hordein ,, ,, Barley (*Hordeum vulgare*).

Zein ,, ,, Maize (*Zea Mays*).

The properties of the gliadins are demonstrated in Expts. 135, 136 and 137).

Glutelins. The proteins of this group are insoluble in water, dilute saline solutions and in alcohol, but they are soluble in dilute alkalies. Glutenin of wheat is the only well-characterized member of this class which has so far been isolated, though other cereals most probably contain similar proteins. A protein of this nature has also been obtained from seeds of Rice (*Oryza sativa*). The properties of the glutelins are demonstrated in Expts. 135 and 136.

CONJUGATED PROTEINS.

Nucleoproteins. Though these proteins probably form constituents of all cells, the only members of the class investigated are those of the wheat embryo. This has been possible since nuclei form a large proportion of the tissue of the embryo. They may be regarded as protein salts of nucleic acid, i.e. protein nucleates. On hydrolysis with acids or enzymes they split up into various proteins and nucleic acid. The nucleoproteins are also connected with the purines (see p. 179).

Nucleic acid. Plant nucleic acids have so far only been investigated from two sources, namely from the embryo of Wheat and from the Yeast cell. These two products appear to be identical, and, on analogy with

animal nucleic acids, it is probable that all plant nucleic acids may prove
to have the same composition. The nucleic acid investigated is a complex
substance formed by the condensation of four nucleotides, each of which
consists of phosphoric acid, a pentose sugar and a purine. Thus yeast
nucleic acid is represented as:

$$\begin{array}{l} \text{HO} \\ \diagdown \\ \text{O}{=}\text{P}-\text{O} \cdot C_5H_7O_2 \cdot C_5H_4N_5O \\ \text{HO}\diagup \text{guanine group} \\ | \\ \text{O} \end{array}$$

$$\begin{array}{l} \text{HO} \\ \diagdown | \\ \text{O}{=}\text{P}-\text{O} \cdot C_5H_6O \cdot C_4H_4N_3O \\ \text{HO}\diagup \text{cytosine group} \\ | \\ \text{O} \end{array}$$

$$\begin{array}{l} \text{HO} \\ \diagdown | \\ \text{O}^-{=}\text{P}-\text{O} \cdot C_5H_6O \cdot C_5H_4N_5 \\ \text{HO}\diagup \text{adenine group} \\ | \\ \text{O} \end{array}$$

$$\begin{array}{l} \text{HO} \\ \diagdown | \\ \text{O}{=}\text{P}-\text{O} \cdot C_5H_7O_2 \cdot C_4H_3N_2O_2 \\ \text{HO}\diagup \text{uracil group} \end{array}$$

On hydrolysis, nucleic acid yields phosphoric acid, d-ribose and the
four purines as ultimate products. Nucleic acid is insoluble in water
but soluble in dilute alkalies: owing to the difficulty of obtaining other
suitable material, nucleic acid is usually prepared from Yeast.

Expt. 131. *Preparation of nucleic acid from Yeast* (from Bertrand, see p. 10).
Take 40 gms. of baker's yeast and add 30 c.c. of 30 % caustic soda solution. Break
up the mass thoroughly and allow it to stand for fifteen minutes. Then add
20 c.c. of water, shake well and at the same time add 10–20 c.c. of 10 % solution of
ferric chloride which will produce a gelatinous precipitate. The mass, which should
be homogeneous, is drained upon a cloth placed in a funnel, so that the almost clear
liquid can be collected in a beaker. The residue is washed with 50 c.c. of warm
water (at 60–70° C.) and again drained on a cloth. The brownish filtrate is added
to an equal volume of alcohol and enough hydrochloric acid is added to render the
mixture slightly acid. A precipitate of nucleic acid is produced. The liquid should
be allowed to stand until the precipitate has settled well. The supernatant fluid
is then decanted, and the precipitate filtered off on a small porcelain funnel using,
if possible, a hardened filter-paper. The precipitate is washed with a little alcohol
and dissolved in the minimum amount of 10 % caustic soda solution. This is reprecipitated by pouring into acid alcohol and finally collected on a small funnel,
again using hardened filter paper.

The nucleic acid is tested for the pentose (ribose) and the phosphoric acid components as follows :

(a) A portion of the precipitate is shaken up with a few c.c. of strong hydrochloric acid in a test tube, a little orcinol is added and the liquid tested for pentoses
(see Expt. 39).

(b) The remainder of the precipitate is boiled for a few minutes with dilute nitric acid (1 part acid : 1 part water) in a test-tube. , Then add an equal volume of 30 % solution of ammonium nitrate and 3–5 drops of concentrated nitric acid. Heat to boiling and add 2 c.c. of a 3 % solution of ammonium molybdate. A yellow precipitate of phosphomolybdate is produced.

DERIVED PROTEINS.

Metaproteins. These are hydrolytic products of albumins and globulins formed by the action of water or dilute acid or alkali. They are insoluble in water, strong mineral acids and all solutions of neutral salts, but are soluble in dilute acids and alkalies in the absence of any large amount of neutral salt.

Expt. 132. *Reactions of metaprotein.* Dissolve about 1 gm. of edestin (see Expt. 139) in 50 c.c. of a 2 % hydrochloric acid and keep on a boiling water-bath for 2 hrs. Neutralize with dilute sodium carbonate solution. A copious precipitate of metaprotein separates out which is insoluble in water. Filter off the precipitate and wash. Make with it the following tests :

(a) Dissolve up some of the precipitate again in 0·4 % hydrochloric acid. To portions of the solution add: (i) Dilute sodium carbonate: the metaprotein is precipitated again and redissolves in excess. (ii) Concentrated hydrochloric acid: the metaprotein is precipitated. (iii) Boil some of the acid solution. No coagulum is formed: the metaprotein is not precipitated by boiling when in solution, and can still be precipitated by neutralizing with sodium carbonate.

(b) Suspend some of the precipitate in water and boil. Cool and add 0·4 % hydrochloric acid: the precipitate is now insoluble, since the metaprotein is coagulated when boiled in suspension.

(c) To some of the precipitate suspended in water, add gradually saturated ammonium sulphate solution: the metaprotein is insoluble in all concentrations of the salt.

Proteoses (albumoses) and peptones. These substances are formed as products of hydrolysis by enzymes. When present in extracts from seeds, however, it is sometimes uncertain whether they formed original constituents of the seeds or resulted from hydrolysis.

As a result of the enzyme hydrolysis of proteins a mixture of several proteoses is usually produced which can be separated by various methods, such as different solubilities in ammonium sulphate, alcohol, etc. The albumoses are soluble in water, salt solutions, dilute acids and alkalies. They are all precipitated by complete saturation with ammonium sulphate, and some by half-saturation with the same salt. On the whole, they give the general colour reactions of the proteins, and are precipitated by the protein precipitants, though some groups of proteoses show certain exceptions. Their solutions are not coagulated on boiling.

The peptones are the only proteins not precipitated by complete saturation with ammonium sulphate. They give the protein colour reactions and are precipitated by tannic acid and lead acetate.

Expt. 133. *Separation and reactions of proteoses.* Prepare about 20 gms. of gluten from 50 gms. of flour as in Expt. 135 (*d*). Put the gluten into a small flask, add 100 c.c. of 0·2 % hydrochloric acid and 0·5 gm. of commercial pepsin: add also a little toluol, shake and plug with cotton-wool. Leave in an incubator at 38° C. for two days. (A control experiment should also be made with 100 c.c. of 0·2 % hydrochloric acid and 0·5 gm. of pepsin. Since pepsin itself gives a biuret reaction, a control is necessary for comparison in the next experiment.) After two days, the incubated mixture is neutralized to litmus with dilute sodium carbonate solution, filtered and saturated while boiling with solid ammonium sulphate. A precipitate of proteoses is formed, which can be gradually collected together as a sticky mass and removed with a glass rod. Dissolve the precipitate in some hot water, filter and make the following tests:

(*a*) *Xanthoproteic reaction.* A positive result is given. A modification of this reaction is characteristic of most proteoses. Add a few drops of nitric acid: a white precipitate is formed which disappears on heating gently and reappears on cooling.

(*b*) *Millon's reaction.* A positive result is given.

(*c*) *Glyoxylic reaction.* A positive result is given.

(*d*) *Biuret reaction.* A pink or pinkish-violet colour is given.

(*e*) *Sulphur reaction.* A positive result is given.

(*f*) Add a little 3 % tannic acid solution. A precipitate is formed.

(*g*) Add a drop of 5 % copper sulphate solution. A precipitate is formed.

(*h*) Add a drop of strong acetic acid and then a couple of drops of 5 % potassium ferrocyanide. A precipitate is formed which disappears on heating gently and reappears on cooling.

(*i*) Boil some of the solution. No coagulum is formed.

Expt. 134. *Detection of peptone.* The saturated solution, from which the proteoses have been precipitated, is then filtered and to a measured quantity (about 5 c.c.) twice the volume of 40 % sodium hydroxide is added and a drop of 1 % copper sulphate solution. A pink colour appears, due to the presence of peptone. A test should be made with the control solution containing hydrochloric acid and pepsin only. An adequate amount should be saturated with ammonium sulphate, filtered and 5 c.c. tested for peptone. The reaction is less marked than in the actual hydrolytic product. Concentrate the remainder of the peptone solution on a water-bath and pour off from the excess of ammonium sulphate crystals. Filter and make the following tests: (i) Xanthoproteic, (ii) Millon's, (iii) Glyoxylic, (iv) Tannic acid. A positive result is obtained in each case.

THE SEED PROTEINS OF CERTAIN PLANTS.

The proteins present in the seeds of certain genera and species, upon which special investigations have been made, may now be considered.

It should be borne in mind that there are always several proteins present in the seed. Some are reserve proteins of the cells of the endosperm or of the storage tissue of the cotyledons: others are proteins of the protoplasm and nuclei of the tissues of the embryo and of the endosperm.

PROTEINS OF CEREALS (GRAMINEAE).

As far as investigations have gone it may be said that the starchy seeds of cereals are poor in albumins and globulins. The chief reserve proteins belong to the peculiar group of prolamins, and a considerable portion also consists of glutelins.

The grain of Wheat (*Triticum vulgare*) contains some proteose and a small percentage of an albumin, leucosin. A globulin occurs only in very small amount. The bulk of the protein consists of gliadin (a prolamin) and of glutenin (a glutelin). Nucleoproteins are present in the embryo, but there is no gliadin or glutenin (Osborne and Voorhees, 16).

The gliadin of wheat has the peculiar property of combining with water to form a sticky mass which binds together the particles of glutenin, the whole forming what is termed gluten. It is this phenomenon which gives the sticky consistency and elastic properties to dough.

Expt. 135. *Extraction of the proteins of the Wheat grain.* (a) *Extraction of albumin (leucosin) and proteose.* Take 100 gms. of white flour (the same quantity of wheat grains which have been ground in a coffee-mill may be used, but the extraction in this case is slower), put the ground mass in a large flask or beaker and add 250 c.c. of distilled water. Let the mixture stand for 1–4 hrs., shaking occasionally. Filter off some of the liquid, first through muslin and then on a filter-pump. Reserve the residue on the filter and test the filtrate for proteins [Expt. 127, (a)–(d)].

Boil a second portion of the filtrate. A precipitate of the albumin, leucosin, is formed. Filter off this precipitate, cool the filtrate and make the protein tests again. All the above tests are given by the proteose in solution. Also make the following special tests for proteoses (Expt. 133). (i) Add a few drops of strong nitric acid. A white precipitate is formed which disappears on heating gently and reappears on cooling. (ii) Add one drop of strong acetic acid and two drops of 5 % potassium ferrocyanide solution. A white precipitate is formed which disappears on heating gently and reappears on cooling.

(b) *Extraction of the globulin.* Take the residue of ground wheat and drain on a filter-pump. Then extract with 250 c.c. of 10 % sodium chloride solution for 12–24 hrs. Filter off, first through muslin, and then through paper on a filter-pump. Put the extract to dialyze in a collodion dialyzer for 24 hrs. (toluol should be added to the liquid in the dialyzer). Filter off the precipitate, which will be very slight, and dissolve it in a little 10 % sodium chloride. (Though so little globulin is present, the experiment is instructive for comparison with the large amount of globulin obtained from many other seeds.) Make the tests for protein [Expt. 127, (a)–(d)] with the solution (Millon's cannot be used on account of the presence of chlorides). Also try the effect of (i) boiling the sodium chloride solution: coagulation is not complete, (ii) adding a little acid: a precipitate is formed as in the case of edestin.

(c) *Extraction of gliadin.* Take 100 gms. of flour (or ground wheat) and add 125 c.c. of 70 % alcohol. Warm on a water-bath and filter. Repeat the process with another 125 c.c. of alcohol. Evaporate the filtrates, which contain gliadin, on a water-bath (or better distil off the alcohol *in vacuo*). When reduced to about half its bulk, take a little of the filtrate and filter. Divide this filtrate into two parts in test-tubes. To one add water : to the other absolute alcohol. A white precipitate of gliadin is formed in each case, since it is insoluble in both water and strong alcohol, though soluble in dilute alcohol. The remainder of the gliadin extract is evaporated almost to dryness, and then poured slowly into a large volume of distilled water. A milky precipitate of gliadin is formed which may be made to settle by adding a little solid sodium chloride and stirring. Filter off the gliadin and dissolve in 10 % acetic acid. With the solution make the tests for protein [Expt. 127, (a)-(d)].

(d) *Extraction of glutenin.* Take 100 gms. of flour, make it into a firm dough with water in an evaporating dish and allow this to stand for half an hour. The dough consists of gluten (gliadin and glutenin) to which the starch adheres. Then put the dough into a piece of fine muslin and knead and wash thoroughly in a stream of water until all the starch is removed. Collect some of the washings in a beaker and to this suspension of starch add a few drops of iodine solution. It will turn a deep blue-black colour. When the starch is completely washed away, an elastic rubbery mass of gluten will remain.

Take about 10 gms. of the gluten, divide it into small pieces and heat it in a flask on a water-bath with small quantities of 70 % alcohol until the extract gives no, or very little, milkiness (due to gliadin) on pouring into water. Decant off the alcohol from the residue of the glutenin, as it can only be filtered with difficulty. Dissolve the glutenin in 0·2 % caustic potash solution. Neutralize a portion of this solution with deci-normal sulphuric acid, drop by drop. A precipitate of glutenin is formed. Test the remainder for proteins [Expt. 127 (a)-(d)].

Expt. 136. *To demonstrate the fact that gluten formation depends on the presence of gliadin.* Repeat Expt. 135 (d) with flour that has been extracted with 70 % alcohol for two or three days. (The alcohol is allowed to stand on the flour in the cold. It is then poured off, and more added, and the process repeated. The flour is now dried again, first in air, then in the steam-oven and finally is ground in a mortar.) No gluten will be formed on account of the absence of gliadin.

In the Barley (*Hordeum vulgare*) grain, small percentages of an albumin, apparently identical with leucosin, and of a globulin, barley edestin, are present, together with some proteose. The main protein is a prolamin, hordein, very similar to, but not identical with, gliadin. There is no well-defined glutelin (Osborne, 11).

In the Rye (*Secale cereale*) grain there are small percentages of proteose, and of leucosin and edestin. The greater part of the protein is gliadin, said to be identical with that in wheat.

In the Maize (*Zea Mays*) grain there is apparently no true albumin, though there is some proteose. There are small quantities of globulin, but the greater part of the protein is a prolamin, termed zein, and a glutenin (Osborne, 12).

Expt. 137. *Extraction of the prolamin, zein, of the Maize grain.* Grind up 100 gms. of maize grains in a coffee-mill, or preferably use maize meal. Add 250 c.c. of hot 70 % alcohol. Filter, and concentrate the filtrate, which contains the zein, on a water-bath (or, better, distil *in vacuo*). Pour a few drops of the concentrated extract into (1) absolute alcohol, (2) distilled water. As in the case of gliadin and hordein, a precipitate of zein will be formed. Then pour the whole extract, after evaporating to a small bulk, into excess of distilled water, and add a little sodium chloride. The precipitate of zein will slowly settle, and can be filtered off. Zein is not readily soluble in acids and alkalies. Hence Millon's and the xanthoproteic tests should be made on the solid material. Zein does not contain the tryptophane nucleus. To demonstrate this, the glyoxylic reaction should be made by shaking up some solid zein in reduced oxalic acid and adding sulphuric acid and mixing. No purple colour is formed.

PROTEINS OF LEGUMINOUS SEEDS (*LEGUMINOSAE*).

In the Leguminosae, which have starchy seeds, the chief reserve proteins, as contrasted with those of cereals, are globulins. The various proteins occurring may be enumerated as:

Legumin. A globulin which forms the chief protein in the seeds of the Broad Bean (*Vicia Faba*), the Pea (*Pisum sativum*), the Lentil (*Ervum Lens*) and the Vetch (*Vicia sativa*). Legumin itself is soluble in water, but occurs as salts which are insoluble in water and soluble in saline solutions. Some portion can be extracted from the seed by water only.

Vicilin. A globulin occuring in smaller quantities than legumin and found only in the Pea, Bean, and Lentil seeds.

Phaseolin. A globulin forming the bulk of the protein of the Kidney Bean (*Phaseolus vulgaris*).

Conglutin. A globulin forming the bulk of the protein in Lupin (*Lupinus luteus*) seeds.

Legumelin. An albumin found in small quantities in the Pea, Broad Bean, Vetch and Lentil.

Phaselin. An albumin found in small quantity in the seeds of the Kidney Bean (*Phaseolus vulgaris*).

Small quantities of proteoses are found in most of the above seeds.

Expt. 138. *Extraction of the proteins of the Pea* (Pisum sativum) (Osborne and Campbell, 13, 14; Osborne and Harris, 15). As we have seen (Expt. 127), a certain amount of protein, including globulin, goes into solution when ground peas are extracted with water. A more complete method of extraction is as follows. Grind in a coffee-mill 20-30 gms. of peas, add to the ground mass 50-60 c.c. of 10 % sodium chloride solution and allow the mixture to stand for 1-2 hrs. Then filter off and saturate the filtrate with solid ammonium sulphate. The globulins, legumin and vicilin, are precipitated out. Filter off the precipitate, and then take up in dilute ammonium sulphate ($\frac{1}{100}$ saturated) and add saturated ammonium sulphate in the proportion of 150 c.c. to every 100 c.c. of the solution ($\frac{6}{10}$ saturation). The legumin is precipitated and can be filtered off. Saturate the filtrate with ammonium sulphate: the vicilin is precipitated and can be filtered off. Dissolve up a little of each precipitate in 10 % sodium chloride, and boil. The vicilin is coagulated, but the legumin is not. Then dissolve up the remainder of the precipitates in dilute ammonium sulphate, and test both the solutions for protein by the usual reactions [Expt. 127, (a)-(d)].

The albumin, legumelin, which occurs only in small quantities in the seeds, can be obtained by dialyzing a water extract of the ground peas until all the globulin is precipitated. On filtering and heating the filtrate, a coagulum of legumelin is formed.

PROTEINS OF FAT-CONTAINING SEEDS.

Of the seeds which contain fat as a reserve material, those investigated have been found, in contrast to the cereals, to contain largely globulin as reserve protein. In many cases these globulins have been obtained in crystalline form after extraction from the plant.

The Hemp-seed (*Cannabis sativa*) contains one of the best-known crystalline globulins, namely edestin. Pure neutral edestin is insoluble in water but soluble in salt solutions. In the presence of acid, however, edestin forms salts which are insoluble in salt solutions Hence a solution of edestin in sodium chloride is precipitated by even small quantities of acids, and, conversely, a solution of edestin in acid is precipitated by small quantities of salt (Osborne, 10).

Expt. 139. *Extraction and crystallization of edestin from Hemp-seed.* Take 50 gms. of hemp-seed and grind in a coffee-mill. Put the ground seed in a large evaporating dish and add 200 c.c. of 5 % sodium chloride solution. Heat with a small flame and stir constantly. A thermometer should be kept in the dish, and the liquid must not rise above 60° C. Filter off, in small quantities at a time, keeping the solution in the dish warm. On cooling, the edestin separates out from the filtrate more or less in crystals. To obtain better crystals, filter off the edestin that has been deposited, and pour the filtrate into a dialyzer; add a little toluol, and suspend the dialyzer in running water. As soon as it is cloudy, examine the dialyzed solution for crystals under the microscope. Add a little 5 % sodium chloride solution to the original precipitate of edestin in the filter. Make with the filtrate the following tests: (i) The tests for proteins [Expt. 127, (a)-(d), except Millon's]. (ii) Boil a little of the solution: it is imperfectly coagulated. (iii) Add a little acid: edestin chloride is precipitated.

In the Castor-oil seed (*Ricinus communis*) there is also present a globulin which can be obtained in a crystalline form by the method of Expt. 139. In addition, there is present an albumin, ricin, which has peculiar toxic properties (Osborne, 10).

A well-crystallized globulin can be obtained from the Linseed (*Linum usitatissimum*)(Osborne, 9,10), and a globulin, excelsin, from the Brazil nut (*Bertholletia excelsa*) (Osborne, 10) also in crystalline or semi-crystalline form. Similar globulins can be extracted from a number of other seeds, i.e. Coconut (*Cocos nucifera*), Sunflower (*Helianthus annuus*), Cotton-seed (*Gossypium herbaceum*), Mustard-seed (*Brassica alba*) and many others. The fat is first removed from the ground seed by either ether or benzene; the residue is then extracted with dilute sodium chloride and the extract dialyzed.

THE AMINO-ACIDS.

There is every reason to believe, since they always arise in hydrolysis of proteins, that amino-acids are universally distributed in the plant. It is, however difficult to isolate and detect them, except in certain special cases, as, for instance, in germinating seeds when a large store of protein is being rapidly hydrolyzed and translocated.

A point of interest in connection with amino-acids is the high percentage of glutaminic acid in many proteins especially those of the Gramineae (35–40 %) and Leguminosae (15–20 %). Moreover, since glutaminic and aspartic acids have two carboxyl groups, only one will be combined in the peptide linkage, the other being free. It appears that the free carboxyl groups of these acids are, even in the protein, combined with ammonia forming an amide, $- CONH_2$. Consequently, when proteins containing a high percentage of glutaminic acid are hydrolyzed they yield a correspondingly high percentage (18–23 %) of "amide nitrogen, as ammonia, compared with other proteins (6–7 %). Moreover, as a result of hydrolysis in the plant itself the respective amides, glutamine and asparagine, are formed and not the free acids.

The following is a short account of the occurrence of some of the amino-acids in the free state (see also p. 134).

Valine has been isolated from seedlings of the Vetch (*Vicia*), Lupin (*Lupinus*) and Kidney Bean (*Phaseolus*). It is present in larger amounts in etiolated seedlings of Lupin than in the green plants.

Leucine is widely distributed. It has been isolated from seedlings of *Vicia*, Vegetable Marrow (*Cucurbita*), *Lupinus*, Pea (*Pisum*) and Goosefoot (*Chenopodium*). It has also been found in *Phaseolus*, Water Ranunculus (*Ranunculus aquatilis*), buds of Horse Chestnut (*Aesculus Hippocastanum*) and in small quantities in Potato tubers and other plants.

Isoleucine has been extracted from seedlings of *Vicia sativa*.

Aspartic acid. The *amide* of this acid, i.e. **asparagine,**

$$CONH_2 \cdot CH_2 \cdot CHNH_2 \cdot COOH$$

is widely distributed in plants. It is present in shoots of Asparagus from which it derives its name. It has also been extracted in very considerable quantities from etiolated seedlings of *Vicia*, Lupin, and from various plants such as Potato, Dahlia, Garden Nasturtium (*Tropaeolum*), *Cucurbita* and Sunflower (*Helianthus*).

Expt. 140. *Preparation of asparagine from shoots of Asparagus* (Asparagus officinalis). Weigh out 500 gms. of shoots of asparagus and pound them in a mortar. Put the mass in a large evaporating dish, add 500 c.c. of water and heat on a water-bath. Squeeze the mass through linen and heat the fluid to boiling in a dish. Filter off the coagulated protein and to the filtrate add tannic acid (to precipitate the remaining proteins, proteoses and peptones) until no more precipitate is formed. Filter and remove any excess of tannic acid by adding a concentrated solution of lead acetate drop by drop. Filter off the precipitate and remove any excess of lead acetate with dilute sulphuric acid. Again filter and finally precipitate the asparagine by adding a concentrated solution of mercuric nitrate (acidify the solution when making with a few drops of nitric acid) until no further precipitate is formed. Filter off the mercury precipitate, suspend it in 150–200 c.c. of water, warm slightly and pass sulphuretted hydrogen through until the precipitate is decomposed. Filter off the mercuric sulphide, and suck air through the solution until it ceases to smell of sulphuretted hydrogen. Neutralize the solution and concentrate on a water-bath to a small bulk. Then add about an equal volume of 98 % alcohol. Crystals of asparagine will separate out. Filter off these on a small conical porcelain funnel, wash with alcohol and dry.

Make a solution of the asparagine (or use the commercial substance) in water and perform the following tests:

(*a*) Add a saturated solution of copper acetate. A blue crystalline precipitate of the copper salt of asparagine separates out. Its appearance is hastened by shaking or rubbing.

(*b*) Boil 2–3 c.c. of the solution with one c.c. of 40 % caustic soda solution. Ammonia is evolved and may be detected by holding red litmus paper in the mouth of the test-tube. Fumes of ammonium chloride will also be formed by introducing a glass rod moistened with strong hydrochloric acid into the tube.

Glutaminic acid. The *amide*, again, of this acid, i.e. **glutamine,**

$$CONH_2 \cdot CH_2 \cdot CH_2 \cdot CHNH_2 \cdot COOH$$

is widely distributed. It has been isolated from seedlings of *Cucurbita*, *Lupinus*, *Helianthus*, Castor-oil plant (*Ricinus*), Spruce Fir (*Picea excelsa*) and a number of Cruciferae.

Expt. 141. *Preparation of glutaminic acid from gluten* (from Cole, see p. 10). Prepare gluten from 100 gms. of flour. This should give about 20 gms. of the dry product. Divide the gluten into small pieces and dissolve it in 150 c.c. of concentrated hydrochloric acid in a round bottomed flask heated on a water-bath. Then add 10 gms. of good blood charcoal (Merck's if possible) and boil on a sand-bath with a reflux condenser for six hours. Filter, and evaporate the filtrate *in vacuo* to about 75 c.c. Put the residue into a narrow cylinder, stand this in ice and saturate with dry hydrochloric acid gas. (This is prepared by slowly dropping strong sulphuric acid from a separating funnel fitted into a flask containing strong hydrochloric acid, and then passing the gas evolved through a second flask of strong sulphuric acid.) Keep the liquid in a cool place for 24 hours, then cool with ice. Crystalline glutaminic hydrochloride will separate out. Add an equal volume of ice-cold alcohol and allow the mixture to stand. Filter on a porcelain funnel through hardened filter-paper or linen. Wash with ice-cold strong hydrochloric acid. Dry in a desiccator over potash and sulphuric acid. Glutaminic acid can be prepared from the hydrochloride by dissolving this in the minimal amount of water and adding 5·3 c.c. of normal caustic soda solution for every gram of product taken. If the solution is then evaporated and cooled, glutaminic acid will separate out.

Arginine has been isolated from seedlings of *Lupinus, Cucurbita, Vicia,* and *Pisum.* It is especially abundant in the seedlings of some Coniferae, i.e. *Picea excelsa,* Silver Fir (*Abies pectinata*) and Scotch Fir (*Pinus sylvestris*). It also occurs in roots and tubers, as for instance in those of the Turnip (*Brassica campestris*), Artichoke (*Helianthus tuberosus*), Chicory (*Cichorium Intybus*), Beet (*Beta vulgaris*), Potato and Dahlia, and in the inner leaves of the Cabbage (*Brassica oleracea*).

Lysine has been isolated from seedlings of *Lupinus, Vicia* and *Pisum.* Also from the inner leaves of the Cabbage and tubers of the Potato.

Phenylalaline has been isolated from seedlings of *Lupinus luteus, Vicia sativa* and *Phaseolus vulgaris.*

Tyrosine is very widely distributed. It is present in seedlings of *Vicia sativa, Cucurbita, Lupinus, Tropaeolum* and tubers of Potato, Turnip, Dahlia, Beet and Celery. Also in berries of Elder (*Sambucus*), in Clover (*Trifolium*), Bamboo (*Bambusa*) shoots and other plants.

Proline has been isolated in very small quantities from etiolated seedlings of *Lupinus albus.*

Histidine has been isolated from seedlings of *Lupinus* and tubers of Potato.

Tryptophane is an important amino-acid and is the one most readily detected on account of the characteristic pink or magenta colour given in its free state with bromine water. The glyoxylic reaction (see p. 136) is given by tryptophane in either the combined state in the protein molecule or in the free state. It has been isolated from seedlings of *Lupinus albus* and *Vicia sativa.*

Dihydroxyphenylalanine. This amino-acid, which contains two hydroxyl groups in the ortho position, has not been detected as a constituent of proteins. It occurs in the free state in all parts of the plant of the Broad Bean (*Vicia Faba*) (Guggenheim, 8) and it has also been found in the Velvet Bean (*Stizolobium*). It readily oxidizes in air and is doubtless responsible for the intense black coloration which appears in all parts of the Broad Bean plant after death of the tissues.

Expt. 142. *Extraction of dihydroxyphenylalanine from the Broad Bean* (Vicia Faba). Take one kilo. of green pods of the bean and put them through a mincing machine. Put the minced mass immediately into boiling water, boil for a few minutes and filter through linen, squeezing the residue thoroughly. Then add lead acetate solution to the filtrate until no further precipitate (consisting of lead compounds of proteins, amino-acids, flavones, etc.) is formed, avoiding an excess of acetate. Filter off and discard this precipitate. Then add ammonia to the filtrate until it is distinctly alkaline to litmus. A yellow precipitate of the lead compound of dihydroxyphenylalanine comes down. Filter, and suspend the precipitate in 500 c.c. of water and pass in sulphuretted hydrogen until the precipitate is decomposed. Filter, and evaporate the filtrate to a small bulk *in vacuo* preferably in a current of carbon dioxide. Crystals of dihydroxyphenylalanine will separate out. Make a solution of the crystals and perform the following test. Add 5% ferric chloride solution. A green colour is formed. Then add a little 1% sodium carbonate solution; the green colour changes to violet.

THE PROTEASES.

We have seen in the previous pages that proteins can be hydrolyzed artificially with the intermediate production of proteoses and peptones, and the final production of a number of amino-acids. There is no doubt that this process of hydrolysis takes place in the living plant, and it is believed that the converse process, the synthesis of these proteins from amino-acids, also takes place in the cell.

There is evidence that this hydrolysis of proteins is catalyzed by certain enzymes which have been termed proteases. On analogy with other enzymes, we may suppose that these enzymes also catalyze the synthesis of the proteins.

It seems highly probable that the proteases are of two types:

1. Pepsin-like enzymes, which catalyze the hydrolysis of proteins to peptones, and, in all probability, the reverse process.

2. Erepsin-like enzymes, which catalyze the hydrolysis of albumoses and peptones to amino-acids, and, in all probability, the reverse process.

We now turn to the evidence for the existence of proteases. In autolysis (see p. 20) the hydrolytic activity of many enzymes is uncontrolled, and in the case of the proteins, the amino-acids are formed as end-products. Amino-acids are rarely present in plants in sufficient quantity to be detected readily, at any rate in small quantities of material, but if the tissues are put to autolyze at temperatures of 38–40° C., the amino-acids then accumulate and can be detected. Of all the amino-acids the one which is most readily identified is tryptophane. If the autolyzed product is boiled, acidified and filtered to remove the remaining proteins, and, to the filtrate, bromine is added, drop by drop, the formation of a pink or purple colour will indicate the presence of free tryptophane, and hence it may be assumed that protein-hydrolysis has taken place. Probably the formation of amino-acids in autolysis is a universal property of plant tissues, for tryptophane has been detected on autolysis of many different parts of plants. Examples are the germinating seeds of the Bean (*Vicia Faba*), Scarlet Runner (*Phaseolus multiflorus*), Pea (*Pisum sativum*), Lupin (*Lupinus hirsutus*) and the Maize (*Zea Mays*): and in ungerminated seeds of the above, though less readily. It is also said to be formed on autolysis of leaves of Spinach (*Spinacia*), Cabbage (*Brassica*), Nasturtium (*Tropaeolum majus*), Scarlet Geranium (*Pelargonium zonale*), Dahlia (*Dahlia variabilis*) and others: also of fruits of Melon (*Cucumis Melo*), Cucumber (*Cucumis sativus*), Banana (*Musa sapientum*), Tomato (*Lycopersicum esculentum*) and others: of bulbs of the Tulip (*Tulipa*), Hyacinth (*Hyacinthus orientalis*) and underground roots of Turnip (*Brassica*), Carrot (*Daucus Carota*) and Beet (*Beta vulgaris*) (Vines, 17–19; Blood, 3; Dean, 5, 6).

Expt. 143. *The formation of tryptophane on autolysis of resting seeds.* Grind up in a coffee-mill 15 gms. of Mustard (*Brassica alba*) seed. Transfer to a flask, and add 100 c.c. of distilled water and about 2 c.c. of toluol. Plug the mouth of the flask with cotton-wool and put in an incubator for 3 days. Then filter off the liquid, boil the filtrate and add a few drops of acetic acid. Filter off any precipitate formed, cool the filtrate and add bromine water *slowly and carefully drop by drop*, shaking well after each drop. A pink or purple colour denotes the presence of tryptophane. Excess of bromine will destroy the colour. Then shake up gently with a little amyl alcohol. The purple colour will be extracted by the amyl alcohol which will rise to the top of the water solution. A control experiment should be made using 10 gms. of seed which has been well boiled with water in an evaporating dish.

It has been assumed that the formation of amino-acids from proteins on autolysis is the outcome of two processes, the hydrolysis of proteins to peptones by pepsins, and the hydrolysis of peptones to amino-acids by erepsins.

The next point to be considered is the possibility of detecting these two classes of enzymes separately. If either the pulp, or water extract, of various plant tissues be added to peptone solution and allowed to incubate at 38° C., tryptophane can be readily detected after a day or two. This has been found to be true for the tissues of many seeds, seedlings, roots, stems, leaves and fruits (such as those already mentioned above and others); the result indicates the wide distribution of an erepsin type of enzyme. The detection of this enzyme is facilitated by the addition of the artificial supply of peptone.

Expt. 144. *The detection of erepsins in plants.*

(a) *In resting seeds.* Grind up 10 gms. of seeds in a coffee-mill, and add 100 c.c. of water, 0·2 gm. of Witte's peptone[1] and a little toluol. Incubate for 2–3 days. The following seeds may be used : Hemp (*Cannabis sativa*), Castor-oil (*Ricinus communis*), Pea (*Pisum sativum*), Scarlet Runner (*Phaseolus multiflorus*), Broad Bean (*Vicia Faba*) and fruit of Wheat (*Triticum vulgare*). Test for tryptophane. Controls should be made in these and the following cases.

(b) *In germinating seeds.* Take 10 germinating peas, pound in a mortar, add 100 c.c. of distilled water, 0·2 gm. of Witte's peptone, and a little toluol. Incubate for 3 days. Test for tryptophane.

(c) *In leaves.* Pound up a small cabbage leaf, add 100 c.c. of water, 0·2 gm. of Witte's peptone and a little toluol. Incubate for 3 days. Test for tryptophane.

(d) *In roots.* Pound up about 20 gms. of fresh carrot root. Add about 100 c.c. of water, 0·2 gm. of Witte's peptone and a little toluol. Incubate for 3 days. Test for tryptophane.

The pepsin type of enzyme is less readily detected. It has long been known that the pitchers of the Pitcher-plant (*Nepenthes*) secrete an enzyme which digests fibrin. A few other cases of protein-digesting enzymes are well known, such as the so-called "bromelin" from the fruit of the Pine-apple (*Ananas sativus*), "cradein" from the latex and fruit of the Fig (*Ficus*) and "papain" from the fruit and leaves of the Papaw tree (*Carica Papaya*). Such enzymes were formerly termed "vegetable trypsins" as they were thought to be of the type of animal trypsin which, alone, hydrolyzes proteins to amino-acids. On analogy with the results of research with other enzymes, it seems likely that "papain," "cradein" and "bromelin" are all mixtures of pepsin and erepsin. In addition to these better known cases, it has also been stated that fibrin is digested by extracts or pounded pulp of the fruits of the Cucumber and the Melon, the "germ" (embryo) of Wheat, the bulbs of Tulip and Hyacinth, the seedlings of the Bean, Pea, Scarlet Runner, Lupin and Maize, and the ungerminated seeds of the Pea, Lupin and Maize. These have also been shown to contain erepsin.

[1] Is prepared from fibrin and consists of a mixture of proteoses and peptone. It is free from tryptophane.

A separation of pepsin from erepsin has been achieved in the case of the seeds of the Hemp (*Cannabis sativa*) by means of the different solubilities of the two enzymes in water and salt solutions.

Expt. 145. The extraction and the separation of the two enzymes, erepsin and pepsin, from Hemp-seed (Cannabis sativa)[1]. Weigh out 50 gms. of hemp-seed, grind it in a coffee-mill and extract with 250 c.c. of 10 % sodium chloride solution. Allow the mixture to stand all night and then filter. Both operations should be carried out at as low a temperature as possible. Measure the filtrate, and add acetic acid to the extent of 0·2 %. A dense precipitate is formed. Filter again, keeping as cool as possible.

The acid filtrate contains the erepsin, but not the pepsin. Measure out 40 c.c. into each of three small flasks, and add the following : (i) 0·2 gm. of Witte's peptone, (ii) the same, only boil the whole solution, (iii) 0·2 gm. of carmine fibrin[2]. Add a little toluol to all three flasks, plug with cotton-wool, and incubate for three to four days. Test for tryptophane in flasks (i) and (ii) ; the first gives a marked reaction, the second little or no reaction. The fibrin in (iii) will remain unaltered.

The precipitate produced by the acetic acid is then washed on the filter twice with 100 c.c. of 10 % sodium chloride solution, containing 0·2 % acetic acid, to remove traces of erepsin. The precipitate is then treated with about 70 c.c. of water, allowed to stand for a time, and then filtered. The filtrate is divided into three equal portions. Add the following respectively : (i) 0·1 gm. of carmine fibrin, (ii) the same, but the solution is boiled, (iii) 0·2 gm. of Witte's peptone. Add a little toluol to all three flasks, plug with cotton-wool and incubate for 3–4 days. The fibrin will be seen to digest slowly in flask (i) : (ii) will show no digestion, and (iii) will give no tryptophane reaction.

REFERENCES

Books

1. **Abderhalden, E.** Biochemisches Handlexikon, iv. Berlin, 1911.
2. **Osborne, T. B.** The Vegetable Proteins. London, 1924. 2nd ed.

Papers

3. **Blood, A. F.** The Erepsin of the Cabbage (*Brassica oleracea*). *J. Biolog. Chem.*, 1910–1911, Vol. 8, pp. 215–225.
4. **Chibnall, A. C.**, and **Schryver, S. B.** Investigations on the Nitrogenous Metabolism of the Higher Plants. Part I. The Isolation of Proteins from Leaves. *Biochem. J.*, 1921, Vol. 15, pp. 60–75.
5. **Dean, A. L.** On Proteolytic Enzymes. I. *Bot. Gaz.*, 1905, Vol. 39, pp. 321–339.

[1] Vines, S. H. *Ann. Bot.*, 1908, Vol. 22, pp. 103–113.
[2] Freshly washed and finely chopped fibrin is placed in carmine solution (1 gm. carmine, 1 c.c. of ammonia, 400 c.c. of water) for 24 hrs. Then strain off and wash in running water till washings are colourless.

6. **Dean, A. L.** On Proteolytic Enzymes. II. *Bot. Gaz.*, 1905, Vol. 40, pp. 121–134.

7. **Fisher, E. R.** Contributions to the Study of the Vegetable Proteases. *Biochem. J.*, 1919, Vol. 13, pp. 124–134.

8. **Guggenheim, M.** Dioxyphenylalanine, eine neue Aminosäure aus *Vicia faba*. *Zs. physiol. Chem.* 1913, Vol. 88, pp. 276–284.

9. **Osborne, T. B.** Proteids of the Flax-seed. *Amer. Chem. J.*, 1892, Vol. 14, pp. 629–661.

10. **Osborne, T. B.** Crystallised Vegetable Proteids. *Amer. Chem. J.*, 1892, Vol. 14, pp. 662–689.

11. **Osborne, T. B.** The Proteids of Barley. *J. Amer. Chem. Soc.*, 1895, Vol. 17, pp. 539–567.

12. **Osborne, T. B.** The Amount and Properties of the Proteids of the Maize Kernel. *J. Amer. Chem. Soc.*, 1897, Vol. 19, pp. 525–532.

13. **Osborne, T. B.,** and **Campbell, G. F.** Proteids of the Pea. *J. Amer. Chem. Soc.*, 1898, Vol. 20, pp. 348–362.

14. **Osborne, T. B.,** and **Campbell, G. F.** The Proteids of the Pea, Lentil, Horse Bean and Vetch. *J. Amer. Chem. Soc.*, 1898, Vol. 20, pp. 410–419.

15. **Osborne, T. B.,** and **Harris, I. F.** The Proteins of the Pea (*Pisum sativum*). *J. Biol. Chem.*, 1907, Vol. 3, pp. 213–217.

16. **Osborne, T. B.,** and **Voorhees, C. G.** The Proteids of the Wheat-Kernel. *Amer. Chem. J.*, 1893, Vol. 15, pp. 392–471.

17. **Vines, S. H.** Tryptophane in Proteolysis. *Ann. Bot.*, 1902, Vol. 16, pp. 1–22.

18. **Vines, S. H.** Proteolytic Enzymes in Plants. I. *Ann. Bot.*, 1903, Vol. 17, pp. 237–264. II. *Ibid.* pp. 597–616.

19. **Vines, S. H.** The Proteases of Plants. I–VII. *Ann. Bot.*, 1904–1910 Vols. 18–24.

CHAPTER X

GLUCOSIDES

ATTENTION has been drawn (Chapters V and VIII) to the fact that in the plant, compounds containing hydroxyl groups often have one or more of these groups replaced by the $C_6H_{11}O_5$— residue of glucose. Such compounds are termed glucosides. The substances in which this substitution most frequently occurs are of the aromatic class, and the glucosides may be regarded, on the whole, as ester-like compounds of carbohydrates with aromatic substances. The non-sugar portion of the glucoside may vary widely in nature, and may be, for instance, an alcohol, aldehyde, acid, phenol, flavone, etc. The sugar constituent is most frequently glucose, but pentosides, galactosides, mannosides and fructosides are also known. Sometimes more than one monosaccharide takes part in the composition of the glucoside. (These various relationships are shown in the accompanying table.) The inclusion of all glucosides in a class is in a sense artificial: the character held in common (with very few exceptions) is that, on boiling with dilute acids, or, by the action of enzymes, hydrolysis takes place, and the glucoside is split up into glucose (or other sugar) and another organic constituent. A number of compounds occurring as glucosides have already been dealt with, for example, the tannins and flavone, flavonol and anthocyan pigments, but, in these cases, the significance of the compounds lies rather in the nature of their non-sugar constituents than in the fact of their being glucosides.

There are, however, a number of glucosides which have been grouped together and are more readily classified in this way than in any other. Some of them, doubtless, have come into prominence as glucosides on account of their association with well-known and specific enzymes, as, for instance, the glucoside amygdalin associated with the enzyme emulsin, and the glucoside sinigrin with the enzyme myrosin.

The hydrolyzing enzymes are by no means always specific, for *in vitro* one particular enzyme may be able to hydrolyze several glucosides. Many glucoside-splitting enzymes have been described, though there is no reason to suppose that each glucoside is only acted upon by an enzyme specific to that glucoside. It is likely moreover that some of the different enzymes described will probably prove to be identical.

In some cases where more than one monosaccharide is attached to the glucoside, the different sugar groups are removed separately by different enzymes (see later, emulsin, p. 160).

The glucosides as a whole (except flavone, flavonol and anthocyan pigments) are colourless crystalline substances. When extracting them from the plant, it is usually necessary to destroy the accompanying enzyme by dropping the material into boiling alcohol or some other reagent (see autolysis, p. 20).

In Chapter v it has already been mentioned that d-glucose exists in two stereoisomeric forms, the α and the β form.

It was also pointed out that the glucosides can be classed either as α- or β-glucosides, according to whether the α or the β form of glucose combines with the non-glucose residue.

```
    H—C—OR                         RO—C—H
      |                                |
    H—C—OH ⟍                        H—C—OH ⟍
      |          ⟍                     |          ⟍
   HO—C—H           O              HO—C—H           O
      |          ⟋                     |          ⟋
    H—C—OH ⟋                        H—C—OH ⟋
      |                                |
    H—C ⟋                            H—C ⟋
      |                                |
    CH₂OH                            CH₂OH
   α-Glucoside                      β-Glucoside
```

Maltose, for instance, is regarded as an α-glucoside of d-glucose. It has been further shown that the enzyme maltase can only hydrolyze α-glucosides, whereas other enzymes, e.g. the prunase component of emulsin, only act on β-glucosides.

The various glucosides considered in detail in this chapter together with some others are grouped under the following headings (Armstrong, 3):

Glucoside	Plant in which commonly found	Products of hydrolysis
		Alcohols
Coniferin	(Coniferae, *Beta*, *Asparagus*, *Scorzonera*)	Glucose + coniferyl alcohol
Populin	(*Populus*)	Glucose + saligenin + benzoic acid
Salicin	(*Salix*, *Populus*)	Glucose + saligenin
Syringin	(*Ligustrum*, *Syringa*, *Jasminum*)	Glucose + syringenin
		Aldehydes
Amygdalin	(*Prunus*, *Pyrus*)	Glucose + benzaldehyde + prussic acid
Dhurrin	(*Sorghum*)	Glucose + parahydroxybenzaldehyde + prussic acid
Linamarin	(*Linum*, *Phaseolus*)	Glucose + acetone + prussic acid

Glucoside	Plant in which commonly found	Products of hydrolysis
		Aldehydes (cont.)
Prulaurasin	(*Prunus*)	Glucose + benzaldehyde + prussic acid
Prunasin	(*Cerasus, Prunus*)	Glucose + benzaldehyde + prussic acid
Sambunigrin	(*Sambucus*)	Glucose + benzaldehyde + prussic acid
Vicianin	(*Vicia*)	Vicianose + benzaldehyde + prussic acid
		Acids
Gaultherin	(*Gaultheria, Spiraea*)	Glucose + methyl salicylate
Strophanthin	(*Strophanthus*)	Mannose + rhamnose + strophanthidin
		Phenols
Arbutin	(Ericaceae)	Glucose + quinol
Hesperidin	(*Citrus*)	Glucose + rhamnose + hesperetin
Naringin	(*Citrus*)	Glucose + rhamnose + narigenin
Phloridzin	(Rosaceae)	Glucose + phloretin
		Coumarin derivatives
Aesculin	(*Aesculus*)	Glucose + aesculetin
Fraxin	(*Fraxinus*)	Glucose + fraxetin
		Mustard-oils
Glucotropaeolin	(*Tropaeolum, Lepidium*)	Glucose + benzyl isothiocyanate + potassium hydrogen sulphate
Sinalbin	(*Brassica alba*)	Glucose + sinapin acid sulphate + acrinylisothiocyanate
Sinigrin	(*Brassica nigra*)	Glucose + allyl isothiocyanate + potassium hydrogen sulphate
		Flavone and flavonol pigments
Apiin	(*Carum*)	Apiose[1] + apigenin
Isoquercitrin	(*Gossypium*)	Glucose + quercetin
Lotusin	(*Lotus*)	Glucose + prussic acid + lotoflavin
Myricitrin	(*Myrica*)	Rhamnose + myricetin
Quercitrin	(*Quercus, Fraxinus, Thea*)	Rhamnose + quercetin
Robinin	(*Robinia*)	Rhamnose + galactose + kaempferol
Rutin	(*Ruta, Capparis, Polygonum*)	Glucose + rhamnose + quercetin
		Anthocyan pigments
Cyanin	(*Centaurea, Rosa*)	Glucose + cyanidin
Delphinin	(*Delphinium*)	Glucose + oxybenzoic acid + delphinidin
Malvin	(*Malva*)	Glucose + malvidin
Oenin	(*Vitis*)	Glucose + oenidin
Peonin	(*Paeonia*)	Sugar + peonidin
Pelargonin	*Pelargonium, Centaurea*)	Glucose + pelargonidin
		Various constituents
Aucubin	(*Aucuba, Plantago*)	Glucose + aucubigenin
Digitalin	(*Digitalis*)	Glucose + digitalose + digitaligenin
Indican	(*Indigofera*)	Glucose + indoxyl

[1] An abnormal sugar, $C_5H_{10}O_5$, containing a branched chain of carbon atoms.

CYANOPHORIC GLUCOSIDES.

The characteristic of these substances is that they yield prussic acid as one of the products of hydrolysis. They are fairly widely distributed: the following list (Greshoff, 15) includes most of the natural orders in which such glucosides occur: Araceae, Asclepiadaceae, Berberidaceae, Bignoniaceae, Caprifoliaceae, Celastraceae, Compositae, Convolvulaceae, Cruciferae, Euphorbiaceae, Gramineae, Leguminosae, Linaceae, Myrtaceae, Oleaceae, Passifloraceae, Ranunculaceae, Rhamnaceae, Rosaceae, Rubiaceae, Rutaceae, Saxifragaceae, Tiliaceae and Urticaceae.

Amygdalin. This is one of the most important of the cyanophoric glucosides. It occurs in the seeds of the bitter Almond (*Prunus Amygdalus*) but it appears to be almost entirely absent from the sweet or cultivated Almond. It also occurs in the seeds of the other species of *Prunus*—the Plum (*P. domestica*), the Peach (*P. Persica*), etc.—of the Apple (*Pyrus Malus*) and the Mountain Ash (*P. Aucuparia*). It occurs sometimes in leaves, flower and bark.

By the action of an enzyme, originally termed emulsin, which occurs in both the bitter and the sweet varieties of Almond, the glucoside is broken up as follows in two stages:

$$C_{20}H_{27}NO_{11} + H_2O = C_6H_{12}O_6 + C_{14}H_{17}NO_6$$
$$\text{mandelonitrile glucoside (prunasin)}$$

$$C_{14}H_{17}NO_6 + H_2O = C_6H_{12}O_6 + HCN + C_6H_5CHO$$
$$\text{benzaldehyde}$$

It should be noted that the sweet Almond contains emulsin although it is almost entirely free from amygdalin.

Emulsin, moreover, has been shown (Armstrong, Armstrong and Horton, 8) to consist of two enzymes, amygdalase and prunase: amygdalase hydrolyzes amygdalin with formation of mandelonitrile glucoside and glucose, whereas prunase hydrolyzes mandelonitrile glucoside (prunasin) with formation of benzaldehyde, prussic acid and glucose. On the basis of these reactions amygdalin is represented as:

```
          ┌─────────O─────────┐                                            C₆H₅
          │                   │                                             │
CH₂OH CH CHOH CHOH CHOH CH · O · CH₂CH CHOH CHOH CHOH CH · O · CH
                             │                              │               │
                             └──────────O──────────┘               CN
```

Prunasin occurs naturally in the Bird Cherry (*Cerasus Padus*), and it is found that prunase may exist in a plant, e.g. Cherry Laurel (*P. Laurocerasus*), which does not contain amygdalase.

Prulaurasin (*laurocerasin*) is a glucoside occurring in the leaves of the Cherry Laurel (*Prunus Laurocerasus*). It has been represented as racemic mandelonitrile glucoside, prunasin being the dextro form.

Sambunigrin is a glucoside occurring in the leaves of the Elder (*Sambucus nigra*). It has been represented as laevo mandelonitrile glucoside.

When tissues containing cyanophoric glucosides and their corresponding enzymes are submitted to autolysis, injury, or the action of chloroform, hydrolysis takes place (see autolysis, p. 20). A rapid method (Mirande, 17; Armstrong, 5) for detecting the prussic acid is to insert paper dipped in a solution of sodium picrate into a tube containing the plant material together with a few drops of chloroform. In the presence of prussic acid the paper becomes first orange and finally brick-red owing to the formation of picramic acid.

In addition to those previously mentioned there are other British plants, the leaves of which give off prussic acid on autolysis (presumably from cyanophoric glucosides), as for example the Columbine (*Aquilegia vulgaris*), Arum (*Arum maculatum*), Hawthorn (*Crataegus Oxyacantha*), Reed Poa (*Glyceria aquatica*), Bird's-foot Trefoil (*Lotus corniculatus*), Alder Buckthorn (*Rhamnus Frangula*), Black and Red Currant and Gooseberry (*Ribes nigrum, R. rubrum, R. Grossularia*), Meadow Rue (*Thalictrum aquilegifolium*) and the Common and Hairy Vetches (*Vicia sativa* and *V. hirsuta*).

It has been shown (Armstrong, 7) that of the species *L. corniculatus* there is a variety (*L. uliginosus*) (taller and growing in moister situations) which does not produce cyanophoric substances and hence does not give off prussic acid on autolysis.

Expt. 146. *Method of detection of cyanophoric glucosides in the plant.* Take three flasks : in one put a whole leaf of the Cherry Laurel (*Prunus Laurocerasus*) : in the second a leaf which has been torn in pieces and then either pricked with a needle or pounded in a mortar : in the third a leaf with a few drops of chloroform. Cork all three flasks, inserting with the corks a strip of sodium picrate paper. (The paper is prepared in the following way : strips of filter-paper are dipped in a 1 % solution of picric acid, are then suspended on a glass rod and allowed to dry in air. Before using, the paper is moistened with 10 % sodium carbonate solution and is suspended in the moist condition just above the material to be examined. In the presence of prussic acid, the paper first becomes orange-yellow, then orange and finally brick-red.) In a short time the paper in the flask containing the leaf and chloroform will turn red : in the flask with the injured leaf, the reddening will take place rather more slowly, whereas in the case of the entire leaf, the paper will remain yellow.

The above experiment may also be carried out, usually with success, on leaves of the Columbine (*Aquilegia vulgaris*), the Arum (*Arum maculatum*) and plants of the Bird's-foot Trefoil (*Lotus corniculatus*) : also with bitter almonds and apple pips, and young shoots of Flax (*Linum perenne*). In the case of the seeds, these may be used crushed, both with and without chloroform, the uninjured seed being used as a control.

Expt. 147. *Preparation of amygdalin.* Weigh out 100 gms. of bitter almonds. Remove the testas by immersing them for a short time in boiling water. Then pound up the almonds well in a mortar and transfer to a flask. Add about 200–300 c.c. of ether and allow the mixture to stand for 2–12 hours. Filter off the ether and extract again with fresh ether. The greater part of the fat will be removed in this way. Then dry the residue from ether and, as rapidly as possible, extract twice or three times with boiling 90–98 % alcohol which removes the amygdalin. The residue, after ether extraction, contains both amygdalin and emulsin, and, if allowed to stand, the emulsin will hydrolyze the amygdalin : hence the necessity for rapid extraction with alcohol. Evaporate the filtered alcoholic extract on a water-bath or, better, distil *in vacuo* to a small bulk. Then add an equal volume of ether and allow the mixture to stand for a time. The amygdalin separates out on standing. Filter off the precipitate, dissolve in a little hot water and allow to crystallize in a desiccator.

Expt. 148. *Preparation of emulsin* (Bourquelot, 10). Weigh out 25 gms. of sweet almonds. (Bitter almonds can also be used. The sweet variety is preferable ; since from them the emulsin can be more readily prepared free from amygdalin.) Plunge them for a moment into boiling water and remove the testas. Pound thoroughly in a mortar, and extract the bulk of the oil with ether as in the last experiment. Then grind up the residue with 50 c.c. of a mixture of equal parts of distilled water and water saturated with chloroform and allow the whole to stand for 24 hours. Filter by means of a filter-pump, and to the filtrate add glacial acetic acid (1 drop to 15 c.c. of the filtrate) whereby the protein is precipitated. Again filter, and to the filtrate add 3–4 times its volume of 96–98 % alcohol. The emulsin is deposited as a white precipitate. Filter off the precipitate and dissolve it in about 100 c.c. of cold distilled water.

Expt. 149. (*a*) *To demonstrate the hydrolysis of amygdalin by emulsin.* Into each of two flasks put 50 c.c. of a 1–3 % solution of amygdalin. To one flask add 25 c.c. of the emulsin solution prepared in the last experiment. To the other flask add 25 c.c. of enzyme solution after it has been *well* boiled, and again boil the mixture after adding the enzyme. Fit each flask with a cork and sodium picrate paper. The paper in the flask containing the unboiled enzyme will rapidly turn red, the control remaining yellow. Unless both the enzyme and the amygdalin solution are well boiled in the case of the control, the paper may show reddening in time on account of traces of prussic acid present in both solutions.

(*b*) *Simplified method for extraction of amygdalin and emulsin, and demonstration of hydrolysis of amygdalin by emulsin.* Take 12 bitter almonds. Remove the testas by immersing them for a short time in boiling water. Then pound up the almonds well in a mortar and transfer to a flask. Add about 50 c.c. of alcohol and heat to boiling on a water-bath. Filter off the extract, and evaporate it to dryness on a water-bath. The residue will contain amygdalin.

Take six sweet almonds and remove the testas as before. Pound in a mortar and transfer to a flask. Add a little ether and allow to stand for a short time. Pour off the ether, and add a little more which should again be poured off. This removes some of the fat and makes extraction of the emulsin easier. Then extract the residue with about 40 c.c. of distilled water and filter. The filtrate contains the enzyme emulsin.

Take 10 c.c. of the emulsin solution, and divide it into two portions in two test-tubes. Boil one *well* (see Expt. 149 *a*), and to both add equal quantities of a water extract of the amygdalin prepared above. Cork the tubes and insert picric paper with the cork in each case.

It has been found, as previously mentioned, that emulsin can hydrolyze other glucosides, as for instance, salicin (see pp. 50, 167). On hydrolysis, salicin splits up into salicylic alcohol (saligenin) and glucose. Salicin, itself, gives no colour with ferric chloride but saligenin gives a violet colour, and by means of this reaction the course of the hydrolysis can be followed.

Expt. 150. To demonstrate the hydrolysis of salicin by emulsin. To 10 c.c. of a 1 % solution of salicin in a test-tube add 10 c.c. of the emulsin solution prepared in Expt. 148 or 149. As a control, boil in a second test-tube another 10 c.c. of the emulsin solution and add 10 c.c. of salicin solution. After about an hour, add to both test-tubes a few c.c. of strong ferric chloride solution. A purple colour will be given in the first test-tube but no colour in the control. The process of hydrolysis will be accelerated by placing the tubes in an incubator.

A modification can be made as follows. A second pair of test-tubes should be prepared as before and to both sufficient ferric chloride should be added to give a faint yellow tinge. The unboiled mixture will gradually acquire a purple colour at ordinary temperature.

Other cyanophoric glucosides are dhurrin, phaseolunatin (linamarin), lotusin and vicianin.

Dhurrin occurs in seedlings of the Great Millet (*Sorghum vulgare*). On hydrolysis it yields glucose, prussic acid and parahydroxybenzaldehyde ($C_6H_4 \cdot OH \cdot CHO$). It is hydrolyzed by emulsin.

Phaseolunatin occurs in seeds of the wild plants of *Phaseolus lunatus* and in seedlings of Flax (*Linum*). It is associated with an enzyme which hydrolyzes it into acetone, glucose and prussic acid.

Lotusin occurs in *Lotus arabicus*. On hydrolysis by an accompanying enzyme (lotase) it gives glucose, prussic acid and a yellow pigment, lotoflavin.

Vicianin occurs in the seeds of a Vetch (*Vicia angustifolia*). It is hydrolyzed by an accompanying enzyme into prussic acid, benzaldehyde and a disaccharide, vicianose.

MUSTARD-OIL GLUCOSIDES.

These are glucosides containing sulphur and they have been found chiefly among the Cruciferae. Sinigrin and sinalbin, the glucosides of mustard, have been most investigated.

Sinigrin. This glucoside occurs in the seed of Black Mustard (*Brassica nigra*) and other species of *Brassica.* Also in the root of the Horse-radish (*Cochlearia Armoracia*). Sinigrin is hydrolyzed by the enzyme, myrosin (Guignard, 16; Spatzier, 18) (which occurs in the plant together with the glucoside), into allyl isothiocyanate, potassium hydrogen sulphate and glucose:

$$C_{10}H_{16}O_9NS_2K + H_2O = C_3H_5NCS + C_6H_{12}O_6 + KHSO_4$$

Expt. 151. *Extraction of sinigrin from Black Mustard.* Weigh out 100 gms. of Black Mustard seed. Grind the seed in a coffee-mill and afterwards pound in a mortar. Heat 175 c.c. of 85 % alcohol to boiling in a flask on a water-bath and add the pounded mustard, and after boiling about $\frac{1}{2}$ hour, filter and press out the alcohol. Then put the dried cake of residue into 300 c.c. of water and allow the mixture to stand for 12 hours. Press out the liquid and after filtering and neutralizing with barium carbonate, concentrate *in vacuo* to a syrup. Then extract with 90 % alcohol and filter. On concentrating and exposing in a crystallizing dish, the sinigrin separates out in white needles.

Sinalbin occurs in the seeds of White Mustard (*Sinapis alba*). By myrosin it is hydrolyzed to *p*-hydroxybenzylisothiocyanate, acid sinapin sulphate and glucose:

$$C_{30}H_{42}O_{15}N_2S_2 + H_2O = C_6H_{12}O_6 + C_7H_7ONCS + C_{16}H_{24}O_5NHSO_4$$

Expt. 152. *Extraction of sinalbin from White Mustard.* Weigh out 100 gms. of White Mustard seed. Grind and pound well and extract the fat with ether. Then extract with twice its weight of 85–90 % alcohol several times and well press out the alcohol. The extract is evaporated to half its bulk and filtered. On cooling, the sinalbin separates out in crystals.

Expt. 153. *Preparation of myrosin.* Weigh out 50 gms. of White Mustard seed and grind in a coffee-mill. Add 100 c.c. of water and allow the mixture to stand for 12 hours. Then filter and allow the filtrate to run into 200 c.c. of 95–98 % alcohol. A white precipitate is formed which contains the myrosin. Filter off the precipitate and wash on the filter with a little ether.

Expt. 154. *Action of myrosin on sinigrin.* Put into two test-tubes equal quantities of a solution of the sinigrin prepared in Expt. 151. Dissolve some of the myrosin prepared in the last experiment in water and divide the solution into two parts. Heat one part to boiling and then add the two portions respectively to the two test-tubes of sinigrin. Plug both test-tubes with cotton-wool. After about $\frac{1}{2}$ hour a strong pungent smell of mustard oil, allyl isothiocyanate, will be detected in the unboiled tube.

A more simple method of demonstrating the action of myrosin is as follows.

Pound about 5 gms. of Black Mustard seed in a mortar and then boil with water. Some mustard oil will be formed before the myrosin is destroyed, so that boiling should be continued until no pungent odour can be detected. Then filter and cool the solution and divide into two parts. To one add some myrosin solution. To the other an equal quantity of boiled enzyme solution. After $\frac{1}{2}$ hour the smell of allyl isothiocyanate should be detected in the unboiled tube.

SAPONINS.

These substances are very widely distributed, being found in the orders: Araliaceae, Caprifoliaceae, Combretaceae, Compositae, Cucurbitaceae, Gramineae, Guttiferae, Lecythidaceae, Leguminosae, Liliaceae, Loganiaceae, Magnoliaceae, Myrtaceae, Oleaceae, Piperaceae, Pittosporaceae, Polemoniaceae, Polygalaceae, Primulaceae, Proteaceae, Ranunculaceae, Rhamnaceae, Rosaceae, Rutaceae, Saxifragaceae, Thymelaeaceae and the majority of the orders of the cohort Centrospermae. On hydrolysis with dilute mineral acids the saponins yield various sugars—glucose, galactose, arabinose, rhamnose—together with other substances termed sapogenins.

The saponins are mostly amorphous substances readily soluble in water (except in a few cases) giving colloidal solutions. These solutions froth on shaking, and with oils and fats they produce very stable emulsions. By virtue of this property they have been used as substitutes for soap. The Soapwort (*Saponaria*) owes its name to the fact that the root contains a saponin.

COUMARIN GLUCOSIDES.

These substances are hydroxy derivatives of coumarin, which itself may be represented as:

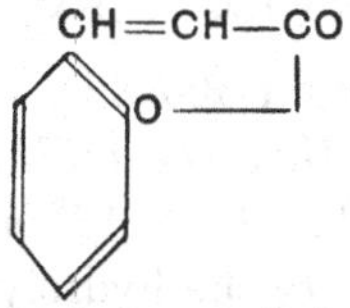

Aesculin is one of the best known of these glucosides. It occurs in the bark of the Horse Chestnut (*Aesculus Hippocastanum*). On hydrolysis with dilute acids it yields glucose and aesculetin, the latter being represented as:

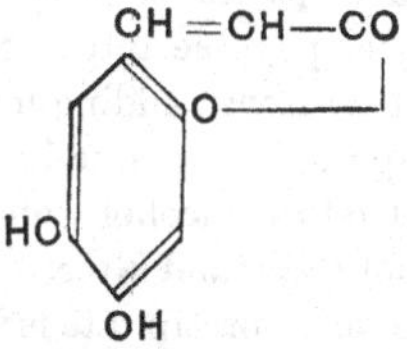

Aesculin is characterized by giving in water solution a blue fluorescence which can be detected even in great dilution. The fluorescence is increased in alkaline, and decreased in acid, solution.

Expt. 155. *Demonstration of the presence of aesculin in* Aesculus *bark.* Strip off the bark from some young twigs of *Aesculus* and boil in a little water in an evaporating dish. Filter and pour the filtrate into excess of water in a large vessel. A blue fluorescent solution will be formed.

GLUCOSIDES OF FLAVONE, FLAVONOL AND ANTHOCYAN PIGMENTS.

These substances have already been considered in Chapter VIII.

GLUCOSIDES OF VARIOUS COMPOSITION.

Coniferin. This glucoside (see also p. 104) occurs in various members of the Coniferae and also in *Asparagus*. On hydrolysis with mineral acids or emulsin, it breaks up as:

$$C_{16}H_{22}O_8 + H_2O = C_6H_{12}O_6 +$$
Coniferin

$$CH = CHCH_2OH$$

OCH$_3$

OH

Coniferyl alcohol

Arbutin. This glucoside is found in the leaves of the Bearberry (*Arctostaphylos Uva-ursi*), *Pyrola*, *Vaccinium*, and other Ericaceae and also of the Pear (*Pyrus communis*).

On hydrolysis with acids arbutin yields quinol and glucose:

$$C_{12}H_{16}O_7 + H_2O = C_6H_6O_2 + C_6H_{12}O_6$$

The same hydrolysis is brought about by the enzyme emulsin.

It has been suggested that the darkening of leaves of the Pear (Bourquelot and Fichtenholz, 11, 12, 13) either on autolysis or injury, or at the fall of the leaf, is due to the hydrolysis of the arbutin by a glucoside-splitting enzyme in the leaf, and subsequent oxidation of the hydroquinone so formed by an oxidase.

Expt. 156. *Extraction of arbutin from leaves of the Pear* (Pyrus communis). Weigh out 100 gms. of fresh leaves (without petioles). Tear the leaves into small pieces and drop them as quickly as possible into about 500 c.c. of boiling 96–98 % alcohol in a flask. Boil for about 20 mins., adding more alcohol if necessary. Then filter off the alcohol and pound up the leaf residue in a mortar and extract again with boiling alcohol. Filter and distil off the alcohol from the extract *in vacuo*. Extract the residue with 100–200 c.c. of hot water and filter. Warm the filtrate and precipitate with lead acetate solution until no more precipitate is formed. This removes flavones, tannins, etc. but the arbutin is not precipitated. Filter and pass sulphuretted hydrogen into the filtrate to remove any excess of lead acetate. Filter and concentrate the filtrate *in vacuo* to a syrup. Then extract twice with small quantities of ethyl acetate. Concentrate the ethyl acetate on a water-bath and cool. A mass of crystals of arbutin will separate out. This should be filtered off on a small filter, and re-crystallized from ethyl acetate. Take up a little of the purified glucoside in water and add a drop or two of ferric chloride solution. A blue coloration will be given.

Salicin. This substance occurs in the bark of various species of Willow (*Salix*) and Poplar (*Populus*): also in the flower-buds of the Meadow-Sweet (*Spiraea Ulmaria*). On hydrolysis with acids, or on treatment with emulsin, salicin is decomposed into saligenin or salicylic alcohol and glucose:

$$C_{13}H_{18}O_7 + H_2O = C_6H_4OH \cdot CH_2OH + C_6H_{12}O_6$$

Saligenin gives a violet colour with ferric chloride solution and in this way the progress of the reaction can be demonstrated (see also p. 163).

Indican. This glucoside occurs in shoots of the so-called "Indigo Plants," *Indigofera Anil, I. erecta, I. tinctoria, I. sumatrana*: also in the Woad (*Isatis tinctoria*), in *Polygonum tinctorium* and species of the Orchids, *Phajus* and *Calanthe*. When boiled with acid or hydrolyzed by an enzyme contained in the plant, it gives glucose and indoxyl:

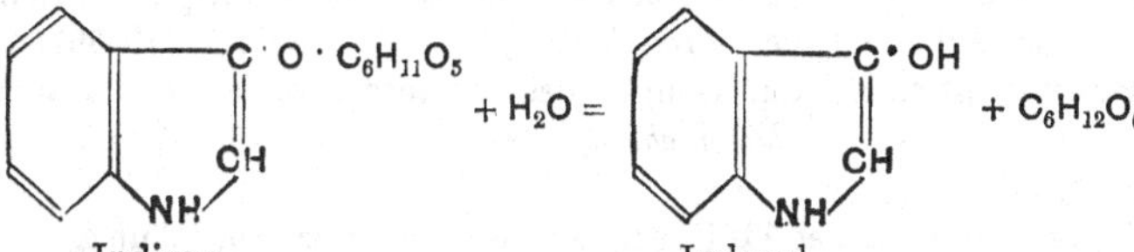

Indican   Indoxyl

The colourless indoxyl can be oxidized either artificially or by an oxidase contained in the plant to a blue product, indigotin or indigo.

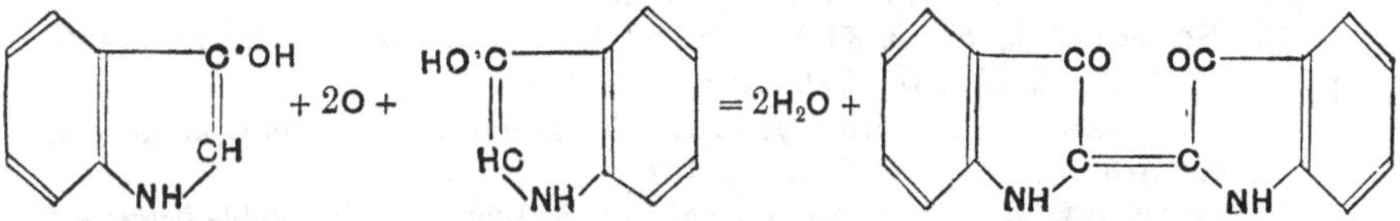

Indoxyl   Indoxyl   Indigo

The relationship between indoxyl and tryptophane (see p. 135) should be noted.

REFERENCES

BOOKS

1. **Abderhalden, E.** Biochemisches Handlexikon, II. Berlin, 1911.
2. **Allen's** Commercial Organic Analysis. Glucosides (E. F. Armstrong), Vol. 7, 1913, pp. 95–135.
3. **Armstrong, E. F.** The Simple Carbohydrates and the Glucosides. London, 1924. 4th ed.
4. **Van Rijn, J. J. L.** Die Glykoside. Berlin, 1900.

PAPERS

5. **Armstrong, E. F.** The Rapid Detection of Emulsin. *J. Physiol.*, 1910, Vol. 40, p. xxxii.
6. **Armstrong, H. E., Armstrong, E. F.**, and **Horton, E.** Studies on Enzyme Action. XII. The Enzymes of Emulsin. *Proc. R. Soc.*, 1908, B Vol. 80, pp. 321–331.
7. **Armstrong, H.E., Armstrong, E.F.**, and **Horton, E.** Herbage Studies. I. *Lotus corniculatus*, a Cyanophoric Plant. *Proc. R. Soc.*, 1912, B Vol. 84, pp. 471–484. II. Variation in *Lotus corniculatus* and *Trifolium repens* (Cyanophoric Plants). *Proc. R. Soc.*, 1913, B Vol. 86, pp. 262–269.
8. **Armstrong, H. E., Armstrong, E. F.**, and **Horton, E.** Studies on Enzyme Action. XVI. The Enzymes of Emulsin. *Proc. R. Soc.*, 1912, B Vol. 85. (i) Prunase, the Correlate of Prunasin, pp. 359–362. (ii) Distribution of β-Enzymes in Plants, pp. 363–369. (iii) Linase and other Enzymes in Linaceae, pp. 370–378.
9. **Armstrong, H. E.**, and **Horton, E.** Studies on Enzyme Action. XIII. Enzymes of the Emulsin Type. *Proc. R. Soc.*, 1910, Vol. 82, pp. 349–367.
10. **Bourquelot, E.** Sur l'emploi des enzymes comme réactifs dans les recherches de laboratoire. *J. pharm. chim.*, 1906, Vol. 24, pp. 165–174 ; 1907, Vol. 25, pp. 16–26, 378–392.
11. **Bourquelot, E.**, et **Fichtenholz, A.** Sur la présence d'un glucoside dans les feuilles de poirier et sur son extraction. *J. pharm. chim.*, 1910, Vol. 2, pp. 97–104.
12. **Bourquelot, E.**, et **Fichtenholz, A.** Nouvelles recherches sur le glucoside des feuilles de poirier : son rôle dans la production des teintes automnales de ces organes. *J. pharm. chim.*, 1911, Vol. 3, pp. 5–13.
13. **Bourquelot, E.**, et **Fichtenholz, A.** Sur le glucoside des feuilles de poirier. *C. R. Acad. sci.*, 1911, Vol. 153, pp. 468–471.
14. **Dunstan, W.**, and **Henry, T. A.** The Chemical Aspects of Cyanogenesis in Plants. *Rep. Brit. Ass.*, 1906, pp. 145–157.
15. **Greshoff, M.** The Distribution of Prussic Acid in the Vegetable Kingdom. *Rep. Brit. Ass.*, 1906, pp. 138–144.
16. **Guignard, L.** Sur quelques propriétés chimiques de la myrosine. *Bul. soc. bot.*, 1894, Vol. 41, pp. 418–428.
17. **Mirande, M.** Influence exercée par certaines vapeurs sur la cyanogenèse végétale. Procédé rapide pour la recherche des plantes à acide cyanhydrique. *C. R. Acad. sci.*, 1909, Vol. 149, pp. 140–142.
18. **Spatzier, W.** Ueber das Auftreten und die physiologische Bedeutung des Myrosins in der Pflanze. *Jahrb. wiss. Bot.*, 1893, Vol. 25, pp. 39–77.
19. **Winterstein, E.**, und **Blau, H.** Beiträge zur Kenntnis der Saponine. *Zs. physiol. Chem.*, 1911, Vol. 75, pp. 410–442.

CHAPTER XI

PLANT BASES

THERE are present in plants a number of substances which form a group, and which may be termed nitrogen bases, or natural bases. These substances are of various constitution but they have the property in common of forming salts with acids by virtue of the presence of primary, secondary, or tertiary amine groupings. Such groupings confer a basic property upon a compound and, as a result, salts are formed with acids on analogy with the formation of ammonium salts:

$$NH_3 + HCl = NH_4Cl \, (NH_3 \cdot HCl)$$

$$CH_3NH_2 + HCl = CH_3NH_2 \cdot HCl$$
methylamine

$$(CH_3)_2 \, NH + HCl = (CH_3)_2 \, NH \cdot HCl$$
dimethylamine

$$(CH_3)_3 \, N + HCl = (CH_3)_3 \, N \cdot HCl$$
trimethylamine

The hydrogen atoms of ammonia can also be replaced by groups of greater complexity, as will be seen below.

Complex ring compounds in which nitrogen forms part of the ring are termed heterocyclic, such as the alkaloids, purines and some amines, for instance pyrrolidine (see below).

The plant bases can be conveniently classified into four groups and this is also to a large extent a natural grouping. They are:

1. Amines } Simpler natural bases.
2. Betaines}
3. Alkaloids.
4. Purine bases.

The first two groups have been termed the simpler natural bases. They are much more widely distributed in the vegetable kingdom than the alkaloids and purines, since they have probably much more significance in general metabolism. The isolation of the simpler bases is a matter of much greater difficulty than that of the alkaloids: the former are soluble in water but insoluble in ether and chloroform, and so are not readily separated from other substances. The alkaloids, however, occur in the plant as salts of acids and if the plant material is made alkaline the free bases can be extracted with ether or chloroform.

The betaines are amino-acids in which the nitrogen atom is completely methylated, and, with one or two exceptions, this grouping does not occur in the true alkaloids. The betaines have only feebly basic properties.

The alkaloids, in contrast to the simpler natural bases, are rather restricted in their distribution, many being limited to a few closely related species or even to one species.

The purine bases are a small group of substances intimately related to each other and to uric acid.

AMINES.

Methylamine, $CH_3 \cdot NH_2$, occurs in the Annual and Perennial Dog's Mercury (*Mercurialis annua* and *M. perennis*) and in the root of the Sweet Flag (*Acorus Calamus*).

Trimethylamine, $(CH_3)_3 \cdot N$, occurs in leaves of the Stinking Goosefoot (*Chenopodium Vulvaria*), in flowers of the Hawthorn (*Crataegus Oxyacantha*) and Mountain Ash (*Pyrus Aucuparia*), and in seeds of *Mercurialis annua*.

Putrescine, $NH_2(CH_2)_4 \cdot NH_2$, occurs in the Thorn Apple (*Datura*) and tetramethylputrescine in a species of Henbane (*Hyoscyamus muticus*).

Hordenine occurs in germinating Barley grains. It is represented as:

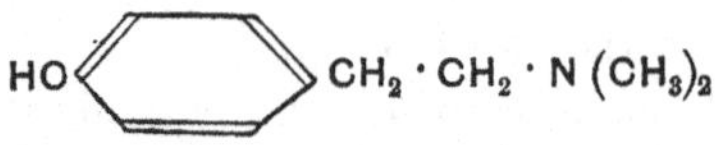

$$HO \langle \quad \rangle CH_2 \cdot CH_2 \cdot N (CH_3)_2$$

Pyrrolidine is said to occur in small quantities in leaves of the Carrot (*Daucus Carota*) and Tobacco (*Nicotiana*) leaves. It is represented as:

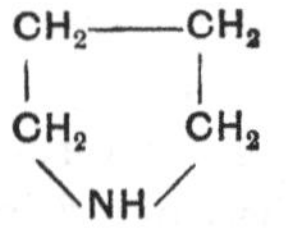

$$\begin{array}{ccc} CH_2 & \!\!\!\!-\!\!\!\!- & CH_2 \\ | & & | \\ CH_2 & & CH_2 \\ & \!\!\searrow\!\! NH \!\!\nearrow\!\! & \end{array}$$

Other amines occur among the lower plants (Fungi).

Choline is sometimes classified with the betaines. It is however intimately connected with lecithin (see p. 98) which is not the case with the betaines. It may be represented as:

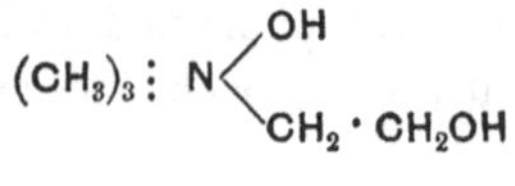

$$(CH_3)_3 : N \Big\langle {}^{OH}_{CH_2 \cdot CH_2OH}$$

Choline is very widely distributed in plants. It is a constituent of the phosphatide, lecithin, and is probably thereby a constituent of all living cells. It has been found in seeds of the Bean (*Vicia Faba*), Pea (*Pisum sativum*), *Strophanthus* spp., Oat (*Avena sativa*), Cotton (*Gossypium herbaceum*), Beech (*Fagus sylvatica*), Fenugreek (*Trigonella Foenum-graecum*) and Hemp (*Cannabis sativa*): in seedlings of Lupins, Soy beans, Barley and Wheat: in Potatoes and Dahlia tubers and in the subterranean parts of Cabbage (*Brassica napus*), Artichoke (*Helianthus tuberosus*), *Scorzonera hispanica*, Chicory (*Cichorium Intybus*), Celery (*Apium graveolens*) and Carrot (*Daucus Carota*); aerial parts of Meadow Sage (*Salvia pratensis*) and Betony (*Betonica officinalis*), and many other tissues. It can only be isolated in very small quantity.

BETAINES.

The betaines, as previously stated, are amino-acids in which the nitrogen atom is completely methylated. Most betaines crystallize with one molecule of water; thus betaine itself in this condition probably has the following constitution, from which its relationship to glycine or aminoacetic acid is indicated:

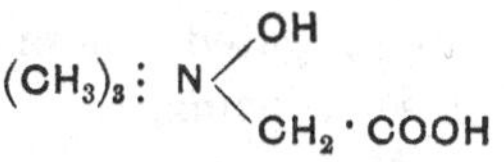

$$(CH_3)_3 : N \Big\langle \begin{matrix} OH \\ CH_2 \cdot COOH \end{matrix} \qquad\qquad H_2N \cdot CH_2 \cdot COOH$$

Betaine or hydroxytrimethyl-   Aminoacetic acid
aminoacetic acid

When dried above 100° C., the betaines lose water and are represented as cyclic anhydrides; thus betaine becomes:

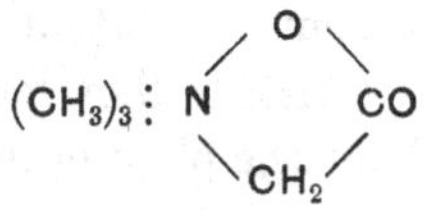

$$(CH_3)_3 : N \begin{matrix} O \\ \diagup \quad \diagdown \\ \qquad CO \\ \diagdown \quad \diagup \\ CH_2 \end{matrix}$$

The individual betaines, probably on account of their close connexion with proteins, are more widely distributed than the individual alkaloids. Further investigation may show an even more general distribution of betaines.

Betaine or trimethylglycine occurs in all species of Chenopodiaceae so far examined including the sugar Beet (*Beta vulgaris*) from which it derives its name: in some genera only of the Amarantaceae: in the "Tea Plant" (*Lycium barbarum*): in seeds of Cotton (*Gossypium herbaceum*), Sunflower (*Helianthus annuus*) and Oat (*Avena sativa*): in tubers of Artichoke (*Helianthus tuberosus*), shoots of Bamboo (*Bambusa*), leaves of Tobacco (*Nicotiana Tabacum*) and in malt and wheat germs.

Stachydrine, though a betaine, is included by most writers among the alkaloids, and this classification has been followed here (see p. 176); it is probably a derivative of proline (see p. 135).

Betonicine, $C_7H_{13}O_3N$, is also, like stachydrine, found in the Betony (*Betonica officinalis*). It is a derivative of oxyproline.

Hypaphorine or trimethyltryptophane, $C_{14}H_{18}O_2N_2$, occurs in the seeds of a tree, *Erythrina Hypaphorus*, which is grown for shade in Coffee plantations.

Trigonelline, like stachydrine, is usually classed with the alkaloids (see p. 175) but it should probably be included among the betaines on account both of its structure and of its wide distribution.

Other betaines, **trimethylhistidine, ergothioneine,** occur in the Fungi.

ALKALOIDS.

The plant alkaloids, so-called because of their basic properties, have attracted considerable attention on account both of their medicinal properties and, in many cases, their intensely poisonous character. They were also the plant bases to be first investigated. As previously mentioned they are not widely distributed, some being, as far as is known, restricted to one genus, or even species. Moreover, several closely related alkaloids are frequently found in the same plant. The orders in which they largely occur are the Apocynaceae, Leguminosae, Papaveraceae, Ranunculaceae, Rubiaceae and Solanaceae.

The alkaloids may be present in solution in the cell-sap in the young tissues, but in older and dead tissues they may occur in the solid state; they may be found throughout the plant or more abundantly in the seed, fruit, root or bark (quinine).

The alkaloids are, as a rule, insoluble in water, but soluble in such reagents as alcohol, ether, chloroform, etc. The majority are crystalline solids which are not volatile without decomposition, but a few, for example coniine, nicotine, which contain no oxygen, are volatile liquids.

The alkaloids occur in the plant as a rule as salts of various organic acids, such as malic, citric, succinic and oxalic, and sometimes with an acid peculiar to the alkaloid with which it is united (e.g. quinic acid in quinine and meconic acid in opium). Artificial salts, i.e. sulphates, chlorides and nitrates, are easily prepared and are readily soluble in water, and from these solutions the free base is precipitated again on addition of alkali.

The alkaloids themselves belong to various classes of compounds, though the basic character always preponderates. Thus, for example, piperine is an amide and can be hydrolyzed into the base piperidine and piperic acid: atropine is an ester made up of the base tropine and tropic acid.

Various methods are employed for the extraction of alkaloids but the exact course of events depends on the alkaloid in question. On the whole the method is either to treat the plant material with alkali and then extract the free alkaloid with ether or chloroform and finally purify by making a salt again; or to extract the alkaloid from the plant with dilute acid, set free the insoluble, or difficultly soluble, base with alkali, and then prepare a salt of the base.

Though individual alkaloids have distinctive reactions, the group as a whole has certain reactions in common, namely the precipitation by the so-called "alkaloidal reagents." These reagents are tannic, phosphotungstic, phosphomolybdic and picric acids, also potassium-mercurio-iodide solution and iodine in potassium iodide solution.

Expt. 157. *General reactions of alkaloids.* Make a 0·5 % solution of quinine sulphate in warm water and add a few drops of each of the following reagents :

(*a*) Tannic acid solution. A white precipitate is formed.

(*b*) Mercuric iodide in potassium iodide solution [Brücke's reagent : 50 gms. of potassium iodide in 500 c.c. water are saturated with mercuric iodide (120 gms.) and made up to 1 litre]. A white precipitate is formed.

(*c*) Phosphotungstic acid (50 gms. of phosphotungstic acid and 30 c.c. of conc. sulphuric acid are dissolved in water and made up to a litre). A white precipitate is formed.

(*d*) Iodine in potassium iodide solution. A brown precipitate is formed.

(*e*) Picric acid solution. A yellow precipitate is formed.

Expt. 158. *Extraction of the free base from quinine sulphate.* Add strong sodium carbonate solution drop by drop to some of the quinine sulphate solution until a white precipitate of quinine is formed. Then add ether and shake up in a separating funnel. The precipitate will disappear as the quinine passes into solution in the ether. Separate off the ethereal solution and let it evaporate in a shallow dish. The quinine is deposited. Take up the quinine again in dilute sulphuric acid and test the solution with the alkaloidal reagents.

The alkaloids are classified into five groups according to the nucleus which constitutes the main structure of the molecule. These five groups are:

1. The pyridine group.
2. The pyrrolidine group.
3. The tropane group.
4. The quinoline group.
5. The isoquinoline group.

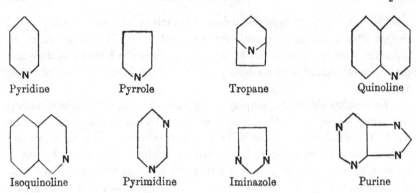

Pyridine   Pyrrole   Tropane   Quinoline

Isoquinoline   Pyrimidine   Iminazole   Purine

1. *The pyridine alkaloids.*

These are, as the name implies, derivatives of pyridine. (Pyridine is a colourless liquid which boils at 115° C. It is a strong base and forms salts with acids.)

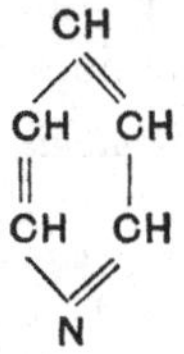

$$
\begin{array}{c}
\text{CH}\\
\diagup\ \diagdown\\
\text{CH}\quad\text{CH}\\
\parallel\quad\ \ \mid\\
\text{CH}\quad\text{CH}\\
\diagdown\ \diagup\\
\text{N}
\end{array}
$$

Pyridine

The more important members of this group are: arecoline, coniine, nicotine, piperine and trigonelline.

Arecoline occurs in the "Betel Nut" which is the fruit of the Areca Palm (*Arecha Catechu*).

Coniine occurs in all parts of the Hemlock (*Conium maculatum*), but more especially in the seed.

Nicotine occurs in the leaves of the Tobacco plant (*Nicotiana Tabacum*). It is a colourless oily liquid which is intensely poisonous. Its constitution may be represented as:

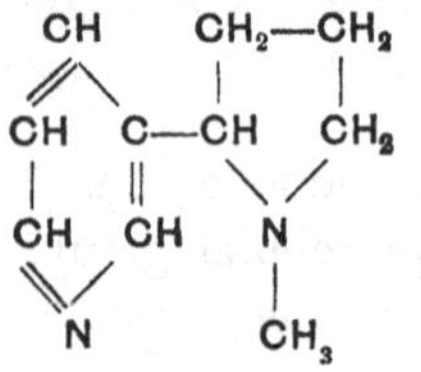

$$
\begin{array}{c}
\text{CH}\qquad\text{CH}_2\!-\!\text{CH}_2\\
\diagup\ \diagdown\qquad\ \mid\qquad\ \ \mid\\
\text{CH}\quad\text{C}\!-\!\text{CH}\quad\text{CH}_2\\
\mid\qquad\parallel\qquad\diagdown\ \diagup\\
\text{CH}\quad\text{CH}\quad\text{N}\\
\diagdown\ \diagup\qquad\quad\mid\\
\text{N}\qquad\qquad\text{CH}_3
\end{array}
$$

It is readily soluble in water and organic solvents.

Expt. 159. *Extraction and reactions of nicotine.* Weigh out 100 gms. of plug tobacco and boil up the compressed leaves with water in an evaporating dish or in a saucepan. Filter off the extract and concentrate on a water-bath. The concentrated solution is made alkaline with lime and distilled from a round-bottomed flask fitted with a condenser, the flask being heated on a sand-bath. The distillate has an unpleasant smell and contains nicotine in solution. Test the solution with the alkaloidal reagents employed in Expt. 157. A precipitate will be obtained in each case.

The nicotine can be obtained from solution in the following way. Acidify the aqueous distillate with oxalic acid and concentrate on a water-bath. Make the concentrated solution alkaline with caustic soda, pour into a separating funnel and shake up with ether. Separate the ethereal extract and distil off the ether. The nicotine is left behind as an oily liquid which oxidizes in air and turns brown. The alkaloidal tests should be made again with the extracted nicotine.

Piperine occurs in various species of Pepper (*Piper nigrum*). The fruit, which is gathered before it is ripe and dried, yields a black pepper, but if the cuticle is first removed by maceration, a white pepper. Piperine is a white solid which is almost insoluble in water but soluble in ether and alcohol.

Expt. 160. *Extraction and reactions of piperine.* Weigh out 100 gms. of black pepper. Put it into an evaporating dish, cover well with lime-water and heat with constant stirring for 15-20 minutes. Then evaporate the mixture completely to dryness on a water-bath. Grind up the residue in a mortar, put it into a thimble and extract with ether in a Soxhlet. Distil off the ether and take up the residue in hot alcohol from which the piperine will crystallize out. With an alcoholic solution make the following tests:

(*a*) Add the alkaloidal reagents mentioned in Expt. 157 and note that a precipitate is formed in each case.

(*b*) Pour a little of the solution into water and note that the piperine is precipitated as a white precipitate.

(*c*) To a little solid piperine in a white dish add some concentrated sulphuric acid. It dissolves to form a deep red solution.

Trigonelline occurs in the seeds of the Fenugreek (*Trigonella Foenum-graecum*), Pea (*Pisum sativum*), Kidney Bean (*Phaseolus vulgaris*), *Strophanthus hispidus*, Hemp (*Cannabis sativa*) and Oat (*Avena sativa*). It is also found in the Coffee Bean (*Coffea arabica*); in tubers of *Stachys tuberifera*, Potato and Dahlia and in roots of *Scorzonera hispanica*. It is really a betaine (see p. 172).

2. *The pyrrolidine alkaloids.*

These are derivatives of pyrrolidine, of which the mother substance is pyrrole. (Pyrrolidine is a liquid boiling at 91° C. It is a strong base and forms stable salts with acids.)

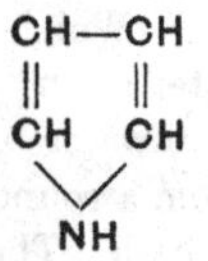

$$\begin{array}{cc} \text{CH---CH} & \text{CH}_2\text{---CH}_2 \\ \parallel \quad \parallel & | \quad\quad | \\ \text{CH} \quad \text{CH} & \text{CH}_2 \quad \text{CH}_2 \\ \diagdown\!\diagup & \diagdown\!\diagup \\ \text{NH} & \text{NH} \\ \text{Pyrrole} & \text{Pyrrolidine} \end{array}$$

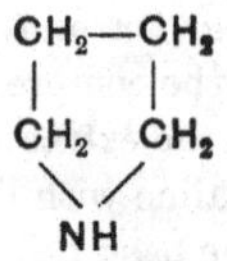

These alkaloids form a small group containing:

Hygrine and **cuskhygrine** which occur in Coca leaves (*Erythroxylon Coca*).

Stachydrine which occurs in tubers of *Stachys tuberifera* and leaves of the Orange Tree (*Citrus Aurantium*) and in various other plants (*Betonica*). The formula is:

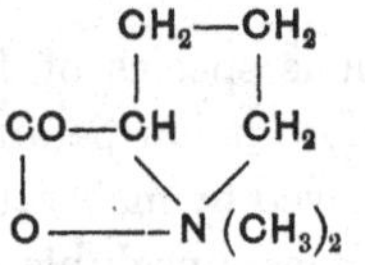

$$\begin{array}{c} \text{CH}_2\text{---CH}_2 \\ | \quad\quad\quad | \\ \text{CO---CH} \quad \text{CH}_2 \\ | \quad\quad \diagdown\!\diagup \\ \text{O------N (CH}_3)_2 \end{array}$$

from which it is seen that it is really a betaine (see p. 172).

3. *The tropane alkaloids.*

These are derivatives of tropane, which may be regarded as formed from condensed piperidine and pyrrolidine groupings. (Tropane is a liquid boiling at 167° C.)

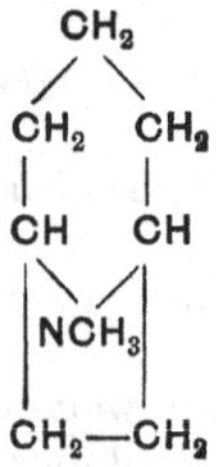

$$\begin{array}{c} \text{CH}_2 \\ \diagup \quad \diagdown \\ \text{CH}_2 \quad \text{CH}_2 \\ | \quad\quad | \\ \text{CH} \quad \text{CH} \\ \diagdown\!\diagup \\ \text{NCH}_3 \\ \diagup\!\diagdown \\ \text{CH}_2\text{---CH}_2 \end{array}$$

Tropane

The alkaloids in this group are limited to four natural orders and are as follows:

Solanaceae: **Atropine** occurs in the root and other parts of the Deadly Nightshade (*Atropa Belladonna*), the Thorn Apple (*Datura Stramonium*) and *Scopolia japonica*. Atropine may be represented as:

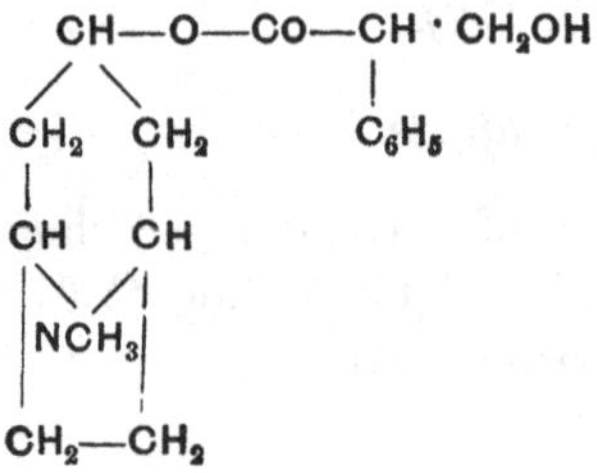

$$\begin{array}{c} \text{CH---O---CO---CH} \cdot \text{CH}_2\text{OH} \\ \diagup \quad \diagdown \quad\quad\quad\quad | \\ \text{CH}_2 \quad \text{CH}_2 \quad\quad \text{C}_6\text{H}_5 \\ | \quad\quad | \\ \text{CH} \quad \text{CH} \\ \diagdown\!\diagup \\ \text{NCH}_3 \\ \diagup\!\diagdown \\ \text{CH}_2\text{---CH}_2 \end{array}$$

Hyoscyamine occurs in the Henbane (*Hyoscyamus niger*), *H. muticus* and also in the Mandrake (*Mandragora*).

Erythroxylaceae: **Cocaine** and **tropacocaine** occur in Coca leaves (*Erythroxylon Coca*) together with smaller quantities of allied alkaloids. Cocaine has the formula:

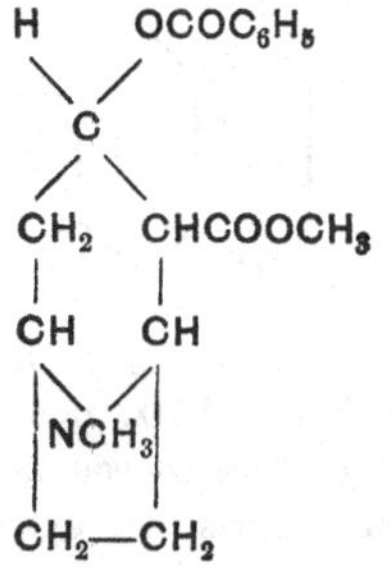

$$H \quad OCOC_6H_5$$

Punicaceae: **Pelletierine** and other allied alkaloids occur in the root and stem of the Pomegranate Tree (*Punica Granatum*).

Leguminosae: **Sparteïne** occurs in the Broom (*Spartium scoparium*): **lupinine** in the yellow and black Lupins (*Lupinus luteus* and *L. niger*) and **cytisine** in the Laburnum (*Cytisus Laburnum*).

4. *The quinoline alkaloids.*

These are derivatives of quinoline. (Quinoline is a colourless liquid which boils at 239° C.) Its constitution is:

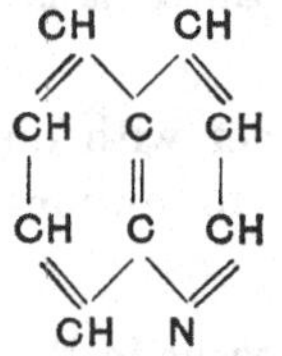

Quinoline

These alkaloids form two natural groups, (*a*) the cinchona alkaloids, i.e. quinine, cinchonine and allied forms, and (*b*) the strychnine alkaloids, i.e. strychnine and brucine.

Quinine occurs in the bark of various species of the genus *Cinchona* (Rubiaceae) which are trees, originally natives of S. America, but now cultivated on a large scale in Ceylon, Java and India. The species employed are *C. Calisaya, Ledgeriana, officinalis, succirubra*. The yellow bark of *Calisaya* has the highest percentage, i.e. 12 %, of alkaloid.

Quinine is a white solid which crystallizes in long needles containing water of crystallization. It is very slightly soluble in cold water, more so in hot but readily soluble in alcohol, ether and chloroform. With acids it forms salts, which are soluble in water, the sulphate being commonly employed in medicine. Quinine is said to have the following constitution:

$$C_{10}H_{15}(OH)N$$

OCH₃

Expt. 161. *Extraction and reactions of quinine.* Mix 20 gms. of quicklime with 200 c.c. of water in a basin and then add 100 gms. of powdered Cinchona bark. Stir together well and then dry the mixture thoroughly on a water-bath, taking care to powder the lumps. The dried mixture is then extracted in a Soxhlet apparatus with chloroform. The chloroform extract is then shaken up in a separating funnel with 25 c.c. of dilute sulphuric acid. The chloroform layer is run off and again extracted with water. The sulphuric acid and water extracts are mixed together and neutralized with ammonia. The liquid is evaporated on a water-bath until crystals of quinine sulphate begin to separate out. With the quinine sulphate the following tests should be made. (It is better to use a solution of the hydrochloride prepared by adding a few drops of hydrochloric acid to the sulphate solution):

(*a*) Test with the alkaloidal reagents of Expt. 157.

(*b*) Add to a little of the solution some bromine water and then some ammonia. A green precipitate is formed which gives a green solution with excess of ammonia.

(*c*) Dissolve a little of the solid quinine sulphate in acetic acid and pour into a large volume of water. A blue opalescence is produced which is characteristic of quinine.

Cinchonine occurs together with quinine in Cinchona bark. It is very similar in constitution to quinine, the latter being methoxy-cinchonine.

Strychnine and **brucine** occur in the seeds of Nux Vomica (*Strychnos Nux-vomica*) and St Ignatius' Bean (*S. Ignatii*).

Expt. 162. *Tests for strychnine.* Add a little concentrated sulphuric acid to a small quantity of strychnine in an evaporating dish and then add a small amount of powdered potassium bichromate. A violet coloration is produced which changes to red and finally yellow.

Curarine, the South American Indian Arrow poison, occurs in several species of *Strychnos* (*S. toxifera* and others).

5. *The isoquinoline alkaloids.*

These can be divided into two groups: (*a*) the opium alkaloids and (*b*) the berberine alkaloids.

The opium alkaloids again fall into two classes: (1) the papaverine group which includes **papaverine, laudanosine, narceine, narcotine** and others, and (2) the morphine group including **morphine, apomorphine, codeine, thebaine** and others.

Opium is the dried latex obtained by making incisions in the capsules of the Opium Poppy (*Papaver somniferum*).

Allied to the papaverine group is **hydrastine** which occurs in the root of *Hydrastis canadensis* (Ranunculaceae).

The constitution of all these alkaloids is very complex.

Expt. 163. *Tests for morphine.*

(*a*) Add a little ferric chloride solution to a solution of a morphine salt. A deep blue coloration is formed.

(*b*) Dissolve some morphine in concentrated sulphuric acid and then after standing about 15 hrs. add concentrated nitric acid. A deep blue-violet colour is produced which afterwards changes to red.

Berberine occurs in the root of the Barberry (*Berberis vulgaris*) and is also found in isolated genera in Anonaceae, Menispermaceae, Papaveraceae, Ranunculaceae and Rutaceae.

Corydaline occurs in *Corydalis cava* (Fumariaceae).

Many other alkaloid substances have been isolated from a large number of different plants, but since the constitution of most of them is unknown, they have not been classified.

PURINE AND PYRIMIDINE BASES.

These substances, as indicated (p. 3), have a hecterocyclic ring structure and are derivatives of purine and pyrimidine: the atoms of the ring are numbered in the order indicated below:

Purine   Pyrimidine

Purine itself is a crystalline basic compound (m. p. 211–212° C.) which forms salts with acids. It is composed of two rings, the pyrimidine and the iminazole: the latter grouping also occurs in histidine (see p. 135).

The chief purine bases which occur in plants are xanthine, guanine, hypoxanthine, adenine, caffeine and theobromine.

Xanthine may be regarded as 2, 6-dioxypurine:

```
        HN——C==O
         |    |
     O==C    C——NH
         |    ‖      \
         |    ‖       CH
         |    ‖      /
        HN——C——N
```

It is widely distributed in plants and has been found in leaves of the Tea plant (*Thea sinensis*), in the sap of the Beetroot (*Beta*) and in various seedlings.

Guanine and **hypoxanthine** can be represented respectively as 2-amino, 6-oxypurine and 6-monoxypurine:

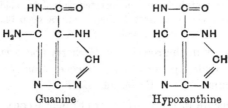

Guanine   Hypoxanthine

They usually occur together and have been found in the germinating seeds of the Sycamore (*Acer pseudoplatanus*), Pumpkin (*Cucurbita Pepo*), Common Vetch (*Vicia sativa*), Meadow Clover (*Trifolium pratense*), yellow Lupin (*Lupinus luteus*) and Barley (*Hordeum vulgare*): also in the juice of the Beet (*Beta*).

Adenine is 6-aminopurine. It is represented as:

```
        N==C——NH₂
         |    |
     HC    C——NH
         ‖    ‖      \
         ‖    ‖       CH
         ‖    ‖      /
         N——C——N
```

It has been found in Beet (*Beta*), Tea leaves (*Thea sinensis*) and in leaves of the Dutch Clover (*Trifolium repens*).

Guanine and adenine are obtained by the hydrolysis of plant nucleoproteins.

Caffeine or **theine** is 1, 3, 7-trimethylxanthine:

```
    CH₃·N——C==O
         |    |
     O==C    C——N·CH₃
         |    ‖      \
         |    ‖       CH
         |    ‖      /
    CH₃·N——C——N
```

It occurs in the leaves and beans of the Coffee plant (*Coffea arabica*), in leaves of the Tea plant (*Thea sinensis*), in leaves of *Ilex paraguensis* ("Paraguay Tea"), in the fruit of *Paullinia Cupana* and in Kola nuts (*Cola acuminata*).

Expt. 164. *Preparation of caffeine from tea*[1]. Digest 100 gms. of tea with 500 c.c. of boiling water for a quarter of an hour. Then filter through thin cloth or fine muslin using a hot-water filter in order to keep the liquid hot. Wash the residue with a further 250 c.c. of boiling water. Add to the filtrate a solution of basic lead acetate until no more precipitate is formed. This removes proteins and tannins. Filter hot and to the boiling filtrate add dilute sulphuric acid until the lead is precipitated as sulphate. Filter from the lead sulphate, and concentrate the solution, with the addition of animal charcoal, to 250–300 c.c. Filter and extract the filtrate three times with small quantities (50 c.c.) of chloroform. Distil off the chloroform on a water-bath, and dissolve the residue in a small quantity of hot water. On allowing the solution to evaporate very slowly, long silky needles of caffeine separate, which may have a slightly yellow tint, in which case they should be drained, redissolved in water, and boiled with the addition of animal charcoal. The yield should be about 1·5 gm.

Evaporate a little of the caffeine on a water-bath with bromine water. A reddish-brown residue is left which becomes purple when treated with ammonia.

Theobromine is 3, 7-dimethylxanthine:

$$
\begin{array}{c}
\text{HN—C=O} \\
\;\;|\;\;\;\;| \\
\text{O=C}\;\;\;\text{C—N·CH}_3 \\
\;\;|\;\;\;\;\|\;\;\;\;\;\backslash \\
\;\;\;\;\;\;\;\;\;\;\;\;\text{CH} \\
\;\;|\;\;\;\;\|\;\;\;\;\;/\!/ \\
\text{CH}_3\text{·N—C—N}
\end{array}
$$

It occurs in the fruit of the Cocoa plant (*Theobroma Cacao*), in leaves of the Tea plant (*Thea sinensis*) and in the Kola nut (*Cola acuminata*).

The chief pyrimidine bases found in the plant are **uracil** (2, 6-dioxy-pyrimidine) and **cytosine** (6-amino-2-oxy-pyrimidine). They are constituents of the molecule of nucleic acid (see p. 141).

It seems appropriate at this point to mention the fact that urea is said to have been detected in small quantity in the Spinach (*Spinacia oleracea*), Cabbage (*Brassica oleracea*), Carrot (*Daucus Carota*), Potato (*Solanum tuberosum*), Chicory (*Cichorium Intybus*) and other plants. A point of considerable interest is the occurrence in the seeds of the Soja Bean (*Glycine hispida*) and other Leguminosae of an enzyme, **urease**, which decomposes urea into ammonia and carbon dioxide:

$$
\text{O=C}\!\!\begin{array}{c}\nearrow\text{NH}_2\\[2pt]\searrow\text{NH}_2\end{array} + \text{H}_2\text{O} = 2\text{NH}_3 + \text{CO}_2.
$$

[1] from Cohen, *Practical Organic Chemistry.*

Urease is quite specific in its action on urea, and the latter has been detected in a few tissues which also yield the enzyme (grain of the Wheat and seeds of the Bean) (Fosse, 4).

Expt. 165. Action of urease on urea. To 100 c.c. of water in a small flask add 1 gm. of urea and 3 gms. of Soja Bean meal. Connect the flask by glass tubing to a second flask containing 0·5 c.c. of strong sulphuric acid in 50 c.c. of water and a piece of litmus paper. Place the flask containing the urea and enzyme in a beaker of water kept at 37–40° C. and run a rapid current of air through the two flasks. After two or three hours, the litmus paper will turn blue. Add sodium carbonate to the second flask and heat. Ammonia will be evolved and can be detected by its smell and by giving white fumes with a drop of strong hydrochloric acid on a glass rod.

REFERENCES

Books

1. **Abderhalden, E.** Biochemisches Handlexikon, v. Berlin, 1911.

2. **Allen's** Commercial Organic Analysis. Vegetable Alkaloids (G. Barger), Vol. 7, 1913, pp. 1–94.

3. **Barger, G.** The simpler Natural Bases. London, 1914.

4. **Fosse, R.** Présence simultanée de l'urée et de l'uréase dans le même végétal. *C. R. Acad. Sci.*, 1914, Vol. 158, pp. 1374–1376.

5. **Henry, T. A.** The Plant Alkaloids. London, 1924. 2nd ed.

6. **Jones, W.** Nucleic Acids. London, 1920. 2nd. ed.

7. **Winterstein, E.,** und **Trier, G.** Die Alkaloide. Berlin, 1910.

APPENDIX

CHAPTER V

PENTOSES

A SUGGESTION as to the origin of pentoses in the plant has been made as follows (Spoehr, see p. 44). It is known that in the case of the hexose sugars, as for instance, glucose and galactose, if the aldehyde group is united to some other group, and thereby prevented from oxidation, the end group —CH_2OH alone may be oxidized under certain conditions with the resultant formation of these two acids respectively:

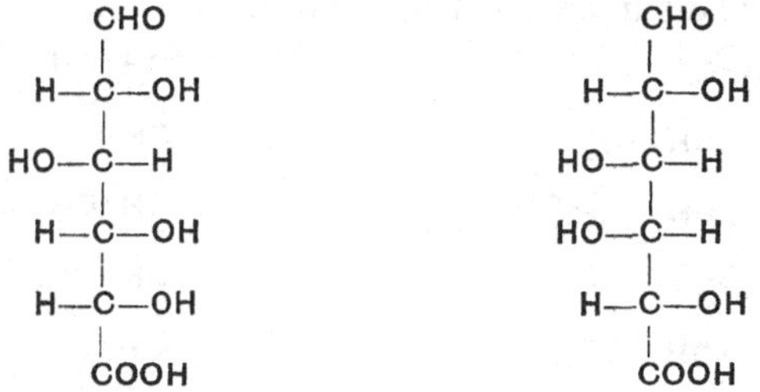

<pre>
        CHO   CHO
         | |
    H—C—OH   H—C—OH
         | |
   HO—C—H   HO—C—H
         | |
    H—C—OH   HO—C—H
         | |
    H—C—OH   H—C—OH
         | |
       COOH   COOH
   Glucuronic acid   Galacturonic acid
</pre>

Glucuronic and galacturonic acids (sometimes, as a type, termed "uronic" acids) are water-soluble, strongly reduce alkaline copper solutions and give the reactions for pentoses (see p. 45). They decompose under certain conditions, giving rise to pentoses and carbon dioxide:

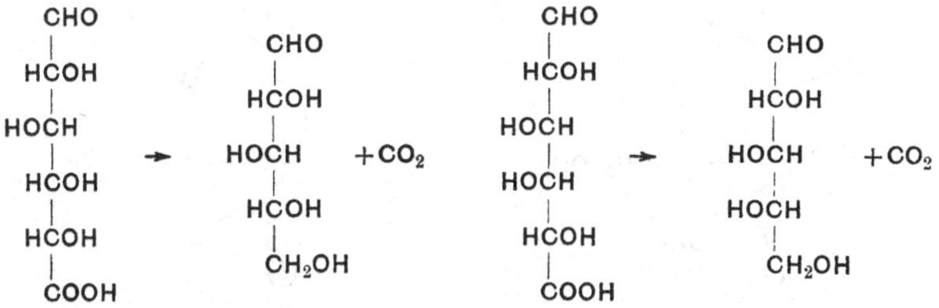

<pre>
   CHO   CHO
    | CHO | CHO
  HCOH | HCOH |
    | HCOH | HCOH
  HOCH | HOCH |
    | →   HOCH   +CO2 | →   HOCH   +CO2
  HCOH | HOCH |
    | HCOH | HOCH
  HCOH | HCOH |
    | CH2OH | CH2OH
  COOH   COOH
 d-Glucuronic acid   d-Xylose   d-Galacturonic acid   l-Arabinose
</pre>

Spoehr has suggested that the above reaction may take place in sunlight. In this connexion it is interesting to note that, in the plant, glucose often occurs associated with xylose, and galactose with ara-

binose (see section on pectic substances below). Spoehr claims to have
isolated glucuronic acid from the plant of *Opuntia* (Cactaceae). Galacturonic acid is a component of the widely-distributed pectic substances.

A method for estimation of the "uronic" acids has been devised
whereby the carbon dioxide evolved on heating with acid is measured.

HEXOSES AND INTERRELATIONSHIPS OF SUGARS.

It has become clear from recent work (Haworth, 4) that in the
normal form of glucose which contains the oxide (lactone) ring, the
linkage is between the first and fifth carbon atoms (amylene oxide)
instead of, as formerly believed, between the first and fourth carbon
atoms (butylene oxide). The butylene oxide probably represents a
specially reactive modification which is known to exist and may be
present in the plant in addition to the ordinary form:

Butylene oxide   Amylene oxide

It is now possible to write the formulae for the hexoses as six-
membered heterocyclic compounds. Expressing the formulae for the
two sugars, *d*-glucose and *d*-galactose thus:

d-Glucose   *d*-Galactose

it will be seen that the two groups (*a*) and (*b*) are interchangeable.
Further, a mechanism has been suggested by which *d*-glucose might
possibly be converted in the plant into *d*-galactose. Let us suppose
that the hydroxyl group (*b*) becomes attached to a residue, such as

phosphoric acid, by condensation with elimination of water. In subsequent hydrolysis of such a compound, a change of position of the groups (a) and (b) is now held to be involved (Walden inversion), and glucose is thereby converted into galactose:

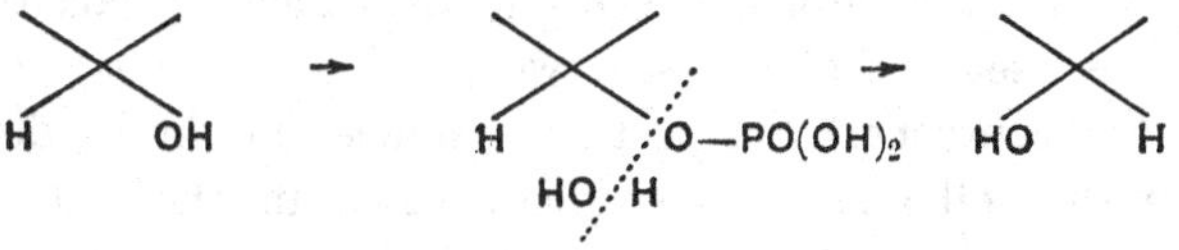

$$\overset{H \qquad OH}{\times} \;\rightarrow\; \underset{HO/H}{\overset{H \qquad O-PO(OH)_2}{\times}} \;\rightarrow\; \overset{HO \qquad H}{\times}$$

A similar inversion might take place involving the conversion of xylose into ribose:

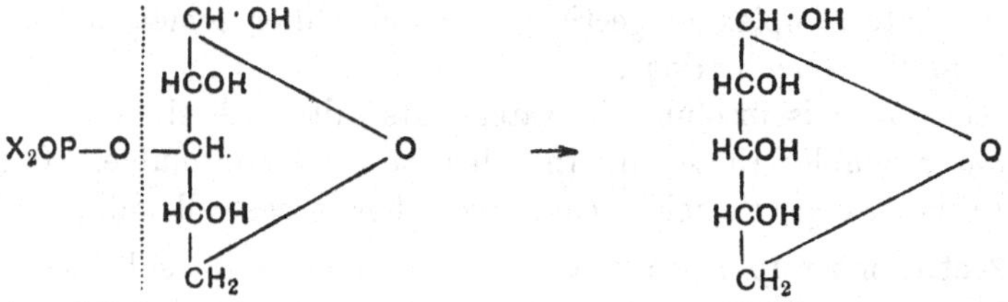

$$X_2OP-O \left|\begin{array}{c} CH \cdot OH \\ | \\ HCOH \\ | \\ -CH. \\ | \\ HCOH \\ | \\ CH_2 \end{array}\right> O \qquad \rightarrow \qquad \left.\begin{array}{c} CH \cdot OH \\ | \\ HCOH \\ | \\ HCOH \\ | \\ HCOH \\ | \\ CH_2 \end{array}\right> O$$

$$d\text{-Xylose} \qquad\qquad\qquad d\text{-Ribose}$$

A point of interest in the above connexion is that both glucose and xylose may be intimately associated with phosphoric acid in the plant; the former in the preliminary stages of respiration (fermentation), the latter in nucleic acid metabolism (see p. 142). Under such conditions, the above interchange might take place. There is, in addition, no other very obvious suggestion for the origin of d-ribose in the plant.

PECTIC SUBSTANCES.

It is probable that the formula suggested by Nanji, Paton and Ling (6) for **pectic acid,** based largely on the results of previous workers, is for the present the most satisfactory. It represents pectic acid as a complex of four molecules of galacturonic acid, one of galactose and one of pentose (arabinose):

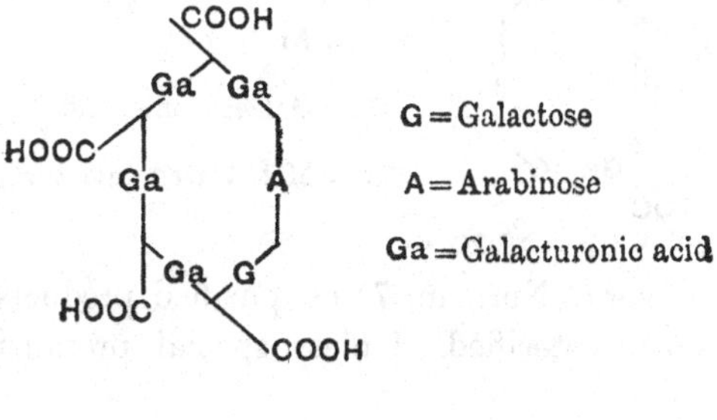

$$
\begin{array}{c}
\text{COOH} \\
\text{Ga}\quad\text{Ga} \\
\text{HOOC}\diagup\qquad\big| \\
\text{Ga}\qquad\text{A} \\
\big|\qquad\diagup \\
\text{Ga}\quad\text{G} \\
\text{HOOC} \\
\text{COOH}
\end{array}
\qquad
\begin{array}{l}
\text{G} = \text{Galactose} \\
\text{A} = \text{Arabinose} \\
\text{Ga} = \text{Galacturonic acid}
\end{array}
$$

Such a formula is in agreement with values, obtained on analysis, of calcium, carbon dioxide and furfural in calcium pectate prepared ·from a number of carefully purified samples of pectic acid from various sources.

From the relationships between glucose, galactose, galacturonic acid and arabinose set out above (see two previous sections), it is readily seen how the components of pectic substances might be derived from glucose in the cell and become incorporated in the cell-wall during development.

That galacturonic acid is the basal unit of pectic substances was first shown by Ehrlich (2). His views, however, on the components of the ultimate complex of pectic compounds differ somewhat in detail from those of other workers.

Pectic acid is insoluble in water. Its salts with the alkali metals are water-soluble: those with the alkaline earths, insoluble. It is uncertain how far pectic acid occurs free either in the cell-wall or cell-sap.

Pectin or soluble pectin is the term given to the substance which is often present in the cell-sap, especially in ripening fruits, and constitutes also, probably, a part of the pectic material of the cell-wall. Pectin may be defined as pectic acid in which some or all of the carboxyl groups are esterified by methyl alcohol. On treatment with alkali, the ester is saponified, and methyl alcohol is set free. The presence of methoxyl groups in pectin was first demonstrated by Fellenberg (3). Like Ehrlich, this author also differs in detail from other investigators as to the components of pectic acid. He, however, considers that in neutral pectin, all carboxyl groups are esterified.

Accepting, for the time being, the above formula of Nanji, Paton and Ling for pectic acid, it is still uncertain how many free carboxyl groups are present in the pectin as it occurs in the plant. According to Norris and Schryver (9) and Norris (8), three groups are usually esterified:

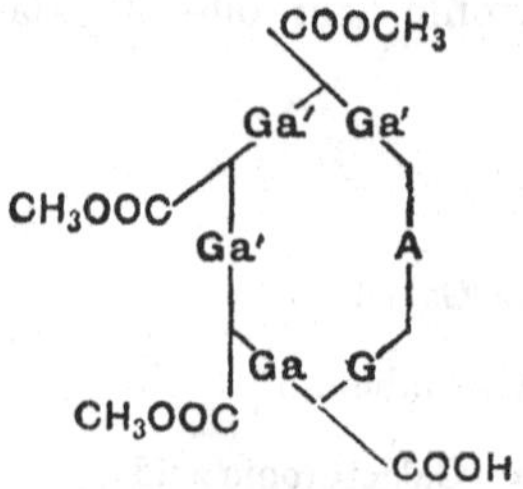

$$COOCH_3$$

CH₃OOC — Ga′ Ga′ Ga′ A

CH₃OOC — Ga G — COOH

G = Galactose

A = Arabinose

Ga = Galacturonic acid

Ga′ = Methylated galacturonic acid

More recent work (Norman, 7) on purified products indicates that all the groups are esterified. Unless special precautions are taken, hydrolysis of the esterified groups is liable to take place during extraction, and this has probably led to varying results in the past.

Insoluble pectin, pectose or **protopectin** is the term given to pectic substances in the cell-wall which are only extracted by more or less prolonged heating with water or acid. It has been held (Fellenberg, 3; Carré, 1) that in insoluble pectin a varying number, sometimes possibly all, of the carboxyl groups of pectic acid are replaced by cellulose groupings with elimination of water, and that the pectic component is only set free on hydrolysis. Nanji and Norman (5), however, suggest that in pectose or protopectin, several molecules of pectin may be associated together, in one or two of which there are some free carboxyl groups, these latter being replaced by iron or calcium. From experience of precipitation of pectin with iron salts, it appears clear that such a compound is only soluble in water, after prolonged treatment at 100° C., but is easily soluble in dilute acid.

Finally, there is yet another type of pectic substance, namely, that in which the pectic acid is combined with a metal, possibly calcium, as **pectate** and is insoluble. This seems to be specially true of the pectic substances of the middle lamella, the first cell-wall formed after cell-division, and eventually the outermost layer of the individual cell-walls. As calcium pectate, the pectic compounds would be readily extracted by ammonium oxalate, or other soluble salt, of which the anion gives an insoluble calcium salt.

From the above short summary, it will be obvious that different pectic compounds will be extracted according to the solvent used. Water will extract soluble pectin. Acid will extract soluble pectin and pectose, whereas ammonium oxalate will extract both of these, in addition to pectic acid occurring either free or as pectate. The product extracted by the method of Schryver and Haynes (see p. 66) is a mixture of more or less de-esterified pectic acids with, possibly, some pectic acid from pectate.

In applying the furfural estimation to pectic acid, it has been realized that this aldehyde may arise from both the pentose and the "uronic" acid components. The latter, therefore, are estimated by the carbon dioxide produced. In performing pentose tests with plant tissues, it should be borne in mind that "uronic" acids may also be responsible for the reactions (see p. 57), since pectic substances are widely, if not universally, present in unlignified tissues.

Pectase. Evidence has been brought forward (Tutin, see p. 80) to show that in the gelatinization of pectin by pectase (see p. 67), the enzyme catalyzes the hydrolysis of the esterified groups in the pectin with the setting free of methyl alcohol and formation of pectic acid.

REFERENCES

1. **Carré, M. H.** Chemical Studies in the Physiology of Apples. IV. Investigations on the Pectic Constituents of Apples. *Ann. Bot.*, 1925, Vol. 39, pp. 811–839.
2. **Ehrlich, F.** Die Pektinstoffe, ihre Konstitution und Bedeutung. *Chem. Z.*, 1917, Vol. 41, pp. 197–200.
3. **Fellenberg, Th. von.** Ueber den Nachweis und die Bestimmung des Methylalkohols, sein Vorkommen in den verschiedenen Nahrungsmitteln und das Verhalten der methylalkoholhaltigen Nahrungsmittel im Organismus. *Biochem. Zs.*, 1918, Vol. 85, pp. 45–117. Ueber die Konstitution der Pektinkörper. *Ibid.*, pp. 118–161.
4. **Haworth, W. N.** Structural Relationships in the Carbohydrate Group. *J. Soc. Chem. Ind.*, 1927, Vol. 46, pp. 295 T–300 T.
5. **Nanji, D. R.,** and **Norman, A. G.** Studies on Pectin. Part II. The Estimation of the Individual Pectic Substances in Nature. *Biochem. J.*, 1928, Vol. 22, pp. 596–604.
6. **Nanji, D. R., Paton, F. J.,** and **Ling, A. R.** Decarboxylation of Polysaccharide Acids: its Application to the Establishment of the Constitution of Pectins and to their Determination. *J. Soc. Chem. Ind.*, 1925, Vol. 44, pp. 253 T–258 T.
7. **Norman, A. G.** Studies on Pectin. Part III. The Degree of Esterification of Pectin in the Juice of the Lemon. *Biochem. J.*, 1928, Vol. 22, pp. 749–752.
8. **Norris, F. W.** The Pectic Substances of Plants. Part IV. The Pectic Substances in the Juice of Oranges. *Biochem. J.*, 1926, Vol. 20, pp. 993–997.
9. **Norris, F. W.,** and **Schryver, S. B.** The Pectic Substances of Plants. Part III. The Nature of Pectinogen and its Relation to Pectic Acid. *Biochem. J.*, 1925, Vol. 19, pp. 676–693.

CHAPTER VII

Lipase. More recent work on *Ricinus* lipase has been carried out by Willstätter and Waldschmidt-Leitz (1). By a special process, these investigators have extracted the enzyme and freed it from crude impurities, though not from accompanying protein. From their researches they draw certain conclusions. In the ungerminated seed, there is a form of lipase (spermatolipase) attached to an insoluble protein: this form is active in acid, but not in neutral, solutions. On germination, spermatolipase gradually disappears and is replaced by another form of lipase (blastolipase) which can hydrolyze fats in neutral or slightly alkaline solution. They maintain that this observed change of enzymatic properties is bound up with a hydrolysis of the protein component of the enzyme and takes place in nature on germination of the seed. This view is corroborated by the fact that the same result is obtained by treatment of spermatolipase with an animal protease. The authors find the effect of acid upon the seed enzyme to be different from that of proteolysis. It confers a power to act subsequently in neutral solution, but the effect is over in a short time. They regard it as due to a simple combination of acid and enzyme.

REFERENCE

1. **Willstätter, R.**, und **Waldschmidt-Leitz, E.** Ueber Ricinuslipase. *Zs. physiol. Chem.*, 1924, Vol. 134, pp. 161–223.

CHAPTER VIII

OXIDIZING ENZYMES

Oxygenase. It has recently been suggested (Szent-Györgyi, 7) that in the oxidation of a compound with the orthodihydroxy grouping of catechol by oxygenase, orthoquinone is formed. In the case of catechol, for instance, the reaction may be represented thus:

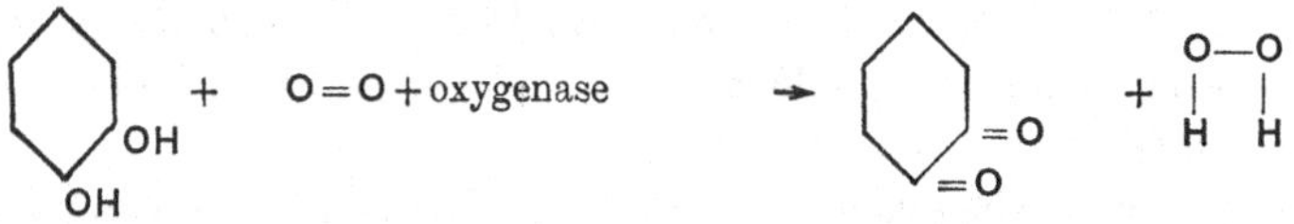

In contrast to the view held by Bach and Chodat that oxidation in such a system involves the taking up of molecular oxygen with formation of peroxide, the reaction represented by the above equation implies that hydrogen is removed from the catechol, molecular oxygen acting as a hydrogen acceptor according to the hypothesis of Wieland.

Orthoquinone is a powerful oxidizing agent and is capable of blueing guaiacum without the intervention of a catalyst[1].

From experimental research on the enzyme system we are considering, certain facts have been ascertained as follows. Hydrogen peroxide has been detected as a product of action of oxygenase (from potato) on

---

[1] It should be noted that orthoquinone, as well as peroxidase and hydrogen peroxide, would be responsible for the blueing obtained on subsequent addition of guaiacum to the products of action of oxygenase on catechol.

catechol (Onslow and Robinson, 2). The test was made with titanium sulphate which gives, as a specific test, a yellow colour with hydrogen peroxide. Further, it seems certain that when oxygenase acts upon catechol, an oxidation product is formed which blues guaiacum in the absence of any catalyst. This has been shown by allowing an enzyme preparation (from potato) to oxidize catechol, and then removing all enzymes by precipitation with colloidal ferric hydroxide. The filtrate, which is free from enzyme, will blue guaiacum. The view that this oxidation product is orthoquinone is placed upon a firmer basis by the isolation of a crystalline anilino-o-quinone (Pugh and Raper, 4) as the result of the action of an oxidizing enzyme (from the meal-worm) on catechol in the presence of aniline. The same crystalline product, moreover, has been obtained by these investigators as a result of the action of peroxidase (from horse-radish) and hydrogen peroxide on catechol:

(ii) catechol (with OH, OH) $+$ $\begin{array}{c} O-O \\ | \quad | \\ H \quad H \end{array}$ $+$ peroxidase $\rightarrow$ o-quinone (=O, =O) $+ 2H_2O$

Even with the above data before us, a difficulty in the interpretation of the mode of action of the catechol-oxygenase system lies in the fact that the oxygenase has not been prepared entirely free from peroxidase. This being so, and as we have experimental proof of the reaction recorded in equation (ii), the possibility arises that orthoquinone may be formed only by the action of peroxidase and hydrogen peroxide on catechol, some other oxidation product and hydrogen peroxide being formed in equation (i).

In any case, the evidence is in favour of orthoquinone being a product of the system, even if the details of the individual reactions are not yet clear. An additional point of interest now arises, in that the orthoquinone thus formed is capable of bringing about certain secondary oxidations, which are, moreover, themselves autocatalytic. A trace of orthoquinone is sufficient to initiate these secondary reactions. Such a trace of orthoquinone (though itself not sufficient to blue guaiacum) is always present adsorbed to the enzyme preparation, however carefully and rapidly this may be prepared. If the usual preparation of oxygenase (see p. 126) is added to a solution of p-cresol or phenol and allowed to stand for a short time, then catechol, as an intermediate product of oxidation, can be detected by means of the colour reaction given with iron salt and dilute alkali (Onslow and Robinson, 1). This secondary oxidation can be represented thus:

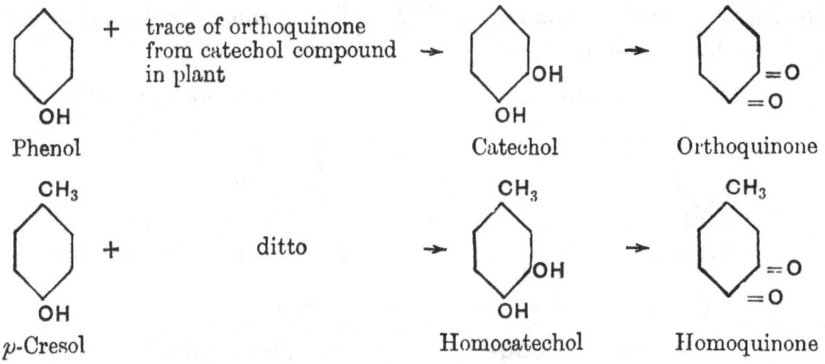

Phenol + trace of orthoquinone from catechol compound in plant → Catechol → Orthoquinone

p-Cresol + ditto → Homocatechol → Homoquinone

From these equations it is obvious that the oxidation of p-cresol or phenol can be initiated by a trace of orthoquinone, and then will proceed autocatalytically. This power of secondary oxidation of phenol and p-cresol can be removed by shaking the solution of the enzyme with animal charcoal which adsorbs and removes the orthoquinone; it can be restored, however, by a trace of catechol which, by providing ortho-quinone, initiates the reaction again (Onslow and Robinson, 3).

Hence the oxidation of p-cresol originally attributed to tyrosinase, is now seen to be a secondary oxidation brought about by the catechol-oxygenase system. All oxygenase preparations from the Higher Plants so far examined have had the same action on p-cresol.

An additional secondary oxidation due, presumably, to orthoquinone, though in this case not autocatalytic, is that of deamination, by oxida-tion, of amino-acids. When oxygenase (from potato) is added to such amino-acids as glycine, alanine, leucine and phenylalanine, there is no oxidation. But, in the presence of catechol, p-cresol or phenol, the amino-acids are oxidized in the following way:

$$CH_2(NH_2)\cdot COOH \quad +O \rightarrow H\cdot CHO + NH_3 + CO_2$$
Glycine

$$CH_3\cdot CH(NH_2)\cdot COOH + O \rightarrow CH_3\cdot CHO + NH_3 + CO_2$$
Alanine

Tyrosinase. It has been stated (p. 129) that no tyrosinase prepara-tion has been obtained from the Higher Plants which does not also act on catechol. It is highly probable that the catechol-oxygenase system and tyrosinase are identical, though this assumption does not at present entirely explain all the observed facts.

The course of the reactions brought about by oxygenase and ortho-quinone (from an orthodihydroxy compound) on tyrosine have been worked out by Raper (5, 6) using an enzyme from the meal-worm.

Dihydroxyphenylalanine (see p. 152) is first formed by the oxidation of tyrosine by orthoquinone:

$CH_2 \cdot CH(NH_2) \cdot COOH$ $CH_2 \cdot CH(NH_2) \cdot COOH$

OH   →   OH / OH

Tyrosine   3·4-Dihydroxyphenylalanine

From this, as usual, an orthoquinone arises:

$CH_2 \cdot CH(NH_2) \cdot COOH$ $CH_2 \cdot CH(NH_2) \cdot COOH$

OH / OH   →   =O / =O

3·4-Dihydroxyphenylalanine   3·4-Quinone of phenylalanine

This undergoes intramolecular change and becomes converted into an indole compound:

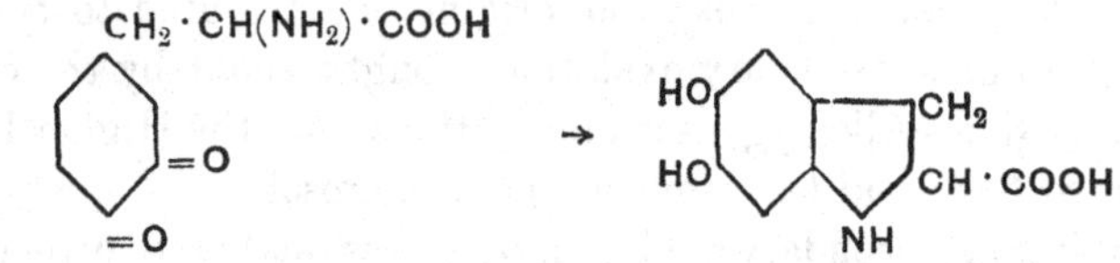

$CH_2 \cdot CH(NH_2) \cdot COOH$ HO / HO — CH_2 / CH·COOH — NH

=O / =O

3·4-Quinone of phenylalanine   5·6-Dihydroxydihydroindole-2-carboxylic acid

which, by subsequent complex changes, is converted into the black product, melanin.

The action of tyrosinase (that is, probably, oxygenase plus orthoquinone) on tyrosine is merely a special case of the secondary oxidation of amino-acids previously described. In tyrosine, the phenolic and amino groupings form parts of the same molecule, and deamination does not follow the usual course.

Peroxidase. Willstätter and his collaborators (8, 9) have evolved a method of preparation and purification of peroxidase whereby samples have been obtained, of which the activity (in terms of capacity to oxidize pyrogallol) is about 5000 times greater than that of the original root. Very briefly, the method is as follows. Dialysable impurities are removed from the root by running water. On treatment with dilute acid, the enzyme is precipitated and adsorbed to the protein, from which it is subsequently extracted (eluted) with dilute alkali. Purification is then carried out by successive adsorptions on to alumina and kaolin, and the enzyme extract is finally concentrated by precipitation. It is free from carbohydrate and protein. The iron content was found to decrease with increased purity of the enzyme.

REFERENCES

1. **Onslow, M. W.**, and **Robinson, M. E.** Oxidising Enzymes. VIII. The Oxidation of certain Parahydroxy-Compounds by Plant Enzymes and its Connection with "Tyrosinase." *Biochem. J.*, 1925, Vol. 19, pp. 420–423.

2. **Onslow, M. W.**, and **Robinson, M. E.** Oxidising Enzymes. IX. On the Mechanism of Plant Oxidases. *Biochem. J.*, 1926, Vol. 20, pp. 1138–1145.

3. **Onslow, M. W.**, and **Robinson, M. E.** Oxidising Enzymes. X. The Relationship of Oxygenase to Tyrosinase. *Biochem. J.*, 1928, Vol. 22, pp. 1327–1331.

4. **Pugh, C. E. M.**, and **Raper, H. S.** The Action of Tyrosinase on Phenols with some Observations on the Classification of Oxidases. *Biochem. J.*, 1927, Vol. 21, 1370–1383.

5. **Raper, H. S.** The Tyrosinase-Tyrosine Reaction. V. Production of $l \cdot 3 \cdot 4$-Dihydroxyphenylalanine from Tyrosine. *Biochem. J.*, 1926, Vol. 20, pp. 735–742.

6. **Raper, H. S.** The Tyrosinase-Tyrosine Reaction. VI. Production from Tyrosine of $5 \cdot 6$-Dihydroxyindole and $5 \cdot 6$-Dihydroxyindole-2-Carboxylic Acid—The Precursors of Melanin. *Biochem. J.*, 1927, Vol. 21, pp. 89–96.

7. **Szent-Györgyi, A. von.** Zellatmung. IV Mitteilung. Ueber den Oxydationsmechanismus der Kartoffeln. *Biochem. Zs.*, 1925, Vol. 162, pp. 399–412.

8. **Willstätter, R.** Ueber Peroxydase. (Zweite Abhandlung.) *Liebigs Ann. Chem.*, 1920, Vol. 422, pp. 47–73.

9. **Willstätter, R.**, und **Pollinger, A.** Ueber Peroxydase. (Dritte Abhandlung.) *Liebigs Ann. Chem.*, 1923, Vol. 430, pp. 269–319.

INDEX

Figures in heavy type denote main references.

Guignard, 164, 168
Gulose, 48
Gum Arabic, 13, 45, 55, 63
—— Tragacanth, 63
Gums, 42, 51, 62
Gun-cotton, 68
Guttiferae, 165

Haas, 10
Harden, 23, 24, 25, 26
Harris, 147, 156
Hatschek, 17
Hawthorn, 111, 113, 125, 161, 170
Haynes, 65, 80
Hazel, 90
—— Nut, 140
Hazel, Red-leaved, 116
Hedge Woundwort, 33
Helianthus, 62, 77, 150
—— *annuus*, 57, 76, 91, 141, 149, 171
—— *tuberosus*, 60, 61, 76, 151, 171
Helleborus niger, 125
Hemerocallis fulva, 76
Hemi-cellulose, 21, 62, 71
Hemlock, 174
Hemp, 69, 90, 139, 140, 171, 175
—— seed, 148, 155
Hemp-nettle, 34
Henbane, 170, 177
Henry, 26, 168, 182
Heracleum, 34, 35
—— *Sphondylium*, 33
Hesperidin, 159
Hesperitin, 159
Hexosephosphatase, 21, 22
Hexosephosphate, 21, 22
Hexoses, 42, 47
Hill, 10
Histidine, 3, 135, 151
—— trimethyl, 172
Hollyhock, 64, 118
Hop, 76
Hopkins, 136
Hordein, 141, 146
Hordenine, 170
Hordeum vulgare, 60, 76, 138, 141, 146, 180
Horse Chestnut, 98, 106, 108, 127, 149, 165
Horse-radish, 124, 125, 126, 164
Horsfall, 112, 130
Horton, 160, 168
Hummel, 113, 130
Humulus Lupulus, 76
Hyacinth, 153, 154
Hyacinthus, 57, 60
—— *orientalis*, 153
Hyaenic acid, 89
Hydrastine, 179
Hydrastis canadensis, 179
Hydrocaffeic acid, 123
Hydrocharis Morsus-ranae, 76
Hygrine, 176
Hymenophyllum demissum, 76
Hyoscyamine, 177
Hyoscyamus muticus, 170, 177
—— *niger*, 177

Hypaphorine, 172
Hypoxanthine, 180

Idaein, 118
Idose, 48
Ilex paraguensis, 181
Iminazole, 3, 174, 179
Indican, 159, 167
Indigo, 167
—— plants, 167
Indigofera, 159
—— *Anil*, 167
—— *erecta*, 167
—— *sumatrana*, 167
—— *tinctoria*, 167
Indole, 129
Indoxyl, 159, 167
Inositol, 21, 101, 102
Inulase, 21, 60
Inulin, 21, 42, 52, 60
—— tests for, 61
Invertase, 21, 22, 25, 52, 78
Invert sugar, 52
Iodoform, 23
Iris, 57, 60
Irvine, 73, 78, 80
Isatis tinctoria, 167
Isobutyric acid, 82
Isocetic acid, 89
Isochlorophyllins, 31
Isoleucine, 134, 150
Isolinolenic acid, 90
Iso-oleic acid, 90
Isoquercitrin, 159
Isoquinoline, 174
Isorhamnose, 42
Isothiocyanate, acrinyl, 159
—— allyl, 21, 159, 164
—— benzyl, 159
—— p-hydroxybenzyl, 164
Isovaleric acid, 82

Jasminum, 55, 158
Jerusalem Artichoke, 60
Jones, 182
Jörgensen, 28, 37, 41
Juglandaceae, 90
Juglans cinerea, 140
—— *nigra*, 140
—— *regia*, 90, 102, 106, 140
Juglansin, 140
Jute, 69

Kaempferol, 113, 120, 121, 159
Kastle, 78, 80
Keracyanin, 118
Kidd, 28, 37, 41
Kidney Bean, 62, 138, 140, 147, 149, 175
Kishida, 110, 131
Kola nut, 181

Labiatae, 109, 123
Laburnum, 177
Laccases, 122, 128
Lacquer, 124

Printed in the United States
By Bookmasters